WOMEN IN MATHEMATICS

RACE, GENDER, AND SCIENCE
Anne Fausto-Sterling, *General Editor*

Feminism and Science
Nancy Tuana, Editor

The "Racial" Economy of Science: Toward a Democratic Future
Sandra Harding, Editor

*The Less Noble Sex: Scientific, Religious, and Philosophical
Conceptions of Woman's Nature*
Nancy Tuana

*Love, Power and Knowledge: Towards a Feminist
Transformation of the Sciences*
Hilary Rose

Women's Health—Missing from U.S. Medicine
Sue V. Rosser

*Deviant Bodies: Critical Perspectives on Difference in
Science and Popular Culture*
Jennifer Terry and Jacqueline Urla, Editors

Im/partial Science: Gender Ideology in Molecular Biology
Bonnie B. Spanier

*Reinventing Biology: Respect for Life and the
Creation of Knowledge*
Lynda Birke and Ruth Hubbard, Editors

Women in Mathematics

The Addition of Difference

Claudia Henrion

INDIANA UNIVERSITY PRESS
Bloomington and Indianapolis

The paper used in this publication meets the
minimum requirements of American National
Standard for Information Sciences—Permanence
of Paper for Printed Library Materials,
ANSI Z39.48-1984.

Manufactured in the United States of America

Library of Congress Cataloging-in-Publication Data
Henrion, Claudia, date.
Women in mathematics : the addition of difference /
Claudia Henrion.
p.   cm. — (Race, gender, and science)
Includes bibliographical references (p. –) and index.
ISBN 0-253-33279-6 (cl : alk. paper).
— ISBN 0-253-21119-0 (pa : alk. paper)
1. Women in mathematics—United States.
2. Women mathematicians—United States—Interviews.
I. Title.   II. Series.
QA27.5.H46   1997
305.43′51—dc21   97-2546

1   2   3   4   5   02   01   00   99   98   97

This book is dedicated to
Dave, Jesse, and Manisha

CONTENTS

ACKNOWLEDGMENTS ix
WOMEN INTERVIEWED xi
INTRODUCTION xvii

1. Rugged Individualism and the Mathematical Marlboro Man 1
   Myth: *Mathematicians work in complete isolation 1*
   Profile: *Karen Uhlenbeck 25*
   Marian Pour-El 49

2. What's a Nice Girl Like You Doing in a Place Like This? 66
   Myth: *Women and mathematics don't mix 66*
   Profile: *Mary Ellen Rudin 85*
   Fan Chung 97

3. Is Mathematics a Young Man's Game? 109
   Myth: *Mathematicians do their best work in their youth 109*
   Profile: *Joan Birman 121*

4. Women and Gender Politics 141
   Myth: *Mathematics and politics don't mix 141*
   Profile: *Lenore Blum 145*
   Judy Roitman 167

5. Double Jeopardy: Gender and Race 188
   Myth: *Only white males do mathematics 188*
   Profile: *Vivienne Malone-Mayes 193*
   Fern Hunt 213

6. The Quest for Certain and Eternal Knowledge 234
   Myths: *Mathematics is a realm of complete objectivity 234*
   Mathematics is non-human 234

Conclusion 263

NOTES 267
RELEVANT LITERATURE 287
INDEX 291

ACKNOWLEDGMENTS

First and foremost I would like to thank the women I interviewed for this book. Without their generosity, their willingness to spend days talking with me, and their openness, this book would not have been possible. My hope is that their stories will help dispel the myth that women can't do mathematics, and simultaneously stimulate dialogue that will make mathematics even more inclusive of women.

I would also like to thank those who provided funding for the early phase of the research: the American Association of University Women, the U.S. Department of Education, the Women's Educational Equity Act, and Middlebury College. I am grateful to my colleagues both inside and outside the mathematics department at Middlebury College and Dartmouth College for their support—in particular Mike Olinick, who wholeheartedly supported this path even though it was an unusual one for a mathematician to be following.

For many mathematicians, myself included, writing is like a second language. I tend to think in terms of visual or symbolic abstractions; hence the actual writing of this book seemed like an ominous task. An old faithful trick that helped me get over the initial hump was to write the manuscript as a series of letters to a childhood friend who is now a philosopher, Alisa Carse. I am eternally grateful for her patience as I sent hundreds of pages of letters to her—my initial attempt to make sense of the thousands of pages of interview transcripts. The book has gone through dozens of revisions, and has been completely recast at least five times. Throughout, Alisa has never faltered in her encouragement. In the last two incarnations of the manuscript, she gave deep and probing feedback that was critical to the final revisions. She also served as a continual reminder that the ideas in this book were of interest to many people (and disciplines) outside mathematics.

A second invaluable reader was Lisa Baker. Our paths intersected at Dartmouth College through our mutual interest in gender and science. A multitalented individual who has since gone into psychology, Lisa read three completely different versions of the manuscript and had many insights early on that were to prove to be prescient. In particular, it was Lisa who first advocated for the integration of the biographies with general themes of the book. Lisa gave copious comments on style, content, and form, all of which made the book what it is today. Her encouragement, optimism, and skillful feedback sustained me through the long dry spells of relative isolation.

There have been many other people who read all or part of the manuscript to whom I am deeply indebted: Anita Solow, Naomi Oreskes, Priscilla Bremser, Annelise Orleck, Alexis Jetter, Ann Bumpus, Susan Brison, Linda Mulley, Viki Kiman. At one critical point when I felt at an impasse, I met with a writer, Jay Stevens, whose path had crossed mine in fortuitous ways. It was he who first asked, "But what is your story behind this story?" And I belligerently resisted: "This is not about my story." Nonetheless, his question was a harbinger of things to come. When I sent the manuscript out to publishers and agents, a literary agent in New York City, Sydelle Kremer, generously spent hours talking with me about the final form the book should take. The questions she asked helped frame the final version, organized around myths of mathematics. She, too, encouraged me to include more of my own story.

In addition, I would like to thank the many scholars and writers whose work formed the foundation that made this book possible. I have included a list of relevant literature at the end of the book, but I would like to mention, in particular, the groundbreaking and thought-provoking work of Evelyn Fox Keller and Sandra Harding. Though their work was often seen as quite controversial in mathematics, it was critical for me in beginning to examine underlying assumptions of the discipline. I am also grateful to Joan Catapano and Jane Lyle at Indiana University Press for their editorial advice and assistance, as well as Anne Fausto-Sterling, who encouraged this project even in its earliest stages, and whose own work was particularly inspiring.

There are many people who played an important role in my life while this book was in process. Their influence, while less obvious, is no less significant. In particular, I would like to thank Ochazania for her many teachings on "the addition of difference," and Viki Kiman for initiating my journey into martial arts. Thanks also to my son, Jesse, who, each evening as I put him to bed, would say, "Tell me about your book, Mama," and to my daughter, Manisha, who is a living reminder that anything is possible.

Finally, the person who deserves more thanks than I can ever begin to offer in this lifetime is my husband, Dave Chapman. With infinite humor, patience, and love, he provided support and encouragement through all phases of the project. Despite my weeks away from home, endless nights working late, and emotional ups and downs, his motto was "Whatever you need to do, do it. We'll make it work."

Joan Birman

Joan Birman is a professor of mathematics at Barnard College and Columbia University. After receiving her B.A. from Barnard, she married, began raising her three children, and held a series of jobs in industry doing applied mathematics. She eventually returned to graduate school, completing her Ph.D. when she was forty-one years old. It is remarkable in mathematics to get a degree so late and then go on to become such a productive and successful mathematician. She was a member of the Institute for Advanced Study at Princeton before beginning her position at Barnard/Columbia, has more than fifty published works in her research area of topology and knot theory, and has given more than a hundred talks in countries around the world, including Japan, Israel, Korea, Denmark, France, and China. Joan Birman exemplifies one way to integrate family with a research career; her life suggests some of the advantages and disadvantages of such a path. Like many of the women in this book, she challenges the myth that "mathematics is a young man's game."

The women described here were interviewed during the period 1988–1993, with updates obtained in 1996. Each interview took place over two to three days and involved between five and ten hours of discussion. The material from the interviews is incorporated throughout the chapters of this book. In addition, in-depth profiles of nine of the eleven women interviewed accompany the first five chapters.

The women interviewed are not meant to be a list of the best and the brightest women mathematicians in the United States. Though certainly some of the top women mathematicians are included, these women were chosen to capture the diverse range of women who are prominent in mathematics; they represent a range of mathematical fields, ages, ethnic and racial backgrounds, geographical locations, and personal and professional situations. There are many other engaging and successful women mathematicians whose stories are well worth hearing. This book is simply one step toward addressing women's invisibility so that there can be no doubt that women can, and do, do mathematics.

Lenore Blum

Lenore Blum has been a full professor of mathematics at Mills College and a research scientist at ICSI (the International Computer Science Institute). From 1992 to 1996 she was the deputy director of the Mathematical Sciences Research Institute in Berkeley. Yet her professional path has not been an easy one. Though she was an extremely promising graduate student from MIT, with strong recommendations from prominent mathematicians in her field, she chose a lecturer position at the University of California at Berkeley over tenure-track jobs at Yale and other institutions, because her husband was also offered a position at Berkeley. At the time she assumed that the position at Berkeley was comparable to the other offers she had. But two years later her position was not renewed. How could she have made such an important misjudgment? How is political wisdom passed down to young mathematicians? What is the role of gender in this process? After leaving Berkeley, Blum secured a position at Mills and became extremely active in promoting women in mathematics, organizing first regional, and later national, conferences and networks for "expanding women's horizons through mathematics." Then, in a remarkable and unusual turn, after years of teaching and administrative work, she returned full-time to mathematics and computer science research—thus defying the myth that once one leaves mathematical research, it is impossible to return.

Fan Chung

The most prolific researcher of this group, Fan Chung has published more than 170 articles in graph theory and discrete mathematics. She is the editor-in-chief of the *Journal of Graph Theory* and is on the editorial board of several other journals. At the time of the interview, Fan was the division manager of mathematics, information sciences, and operations research at Bell Communications Research, thus providing a non-academic perspective on mathematical research. She is now a professor of mathematics at the University of Pennsylvania. Fan was fortunate to have a community of women peers during her undergraduate training in mathematics at Taiwan University. Most of these women, including Fan, came to the United States for graduate work, remained, and went on to become first-rate researchers. She discusses the importance of such a community for her mathematical development. In addition, Fan is married to, and does extensive research with, one of the top mathematicians in the country. How do they integrate their personal and professional relationship?

Marcia Groszek

Marcia Groszek was an undergraduate at Hampshire College and went on to get her Ph.D. at Harvard University. She is now an associate professor at Dartmouth College, where she was for many years the only woman in the mathematics department. How did being the only woman affect her experiences there? As one of the youngest women interviewed, she presents insight into contemporary decisions about combining children with a life as a research mathematician, and the conflicts that arise. She was a participant and an instructor in the Hampshire Summer Math Program for talented high school students. Her research is in logic.

Fern Hunt

At the time of our interview, Fern Hunt was a professor of mathematics at Howard University; she is now a researcher at the National Institute for Standards in Technology. She did her undergraduate work at Bryn Mawr and graduate work at the Courant Institute at New York University. She does research in applied mathematics and mathematical biology, and has had research support from NSF, the National Bureau of Standards, NASA, and the National Institutes of Health. Although Fern knew early on that she was interested in mathematics, she did not consider it a possibility for herself because she was black and a woman. But a black science teacher in high school changed her mind. He was the first person to encourage her in the pursuit of science, both practically, by telling her about programs and schools she should go to, and emotionally, by nurturing her confidence in herself as a talented and gifted young black woman.

Linda Keen

Linda Keen is a full professor of mathematics at Lehman College and a doctoral faculty member in both mathematics and computer science at the Graduate Center of City University of New York. She received her undergraduate degree from CUNY in 1960, and her Ph.D. from NYU in 1964. Her research is in complex analysis and dynamical systems. She has served as a president of the Association for Women in Mathematics, and on numerous committees of professional organizations, including the AMS, MAA, NRC, and NSF. She also raised two children during her professional career and has become acutely aware of the powerful role that personal and professional support play in one's professional development. Like many women in the book, she feels that her best work has been done later in her life; now that her children are older and independent, she can apply a single-minded focus to mathematics.

Vivienne Malone-Mayes

Vivienne Malone-Mayes was a professor of mathematics at Baylor University. She received her B.A. and M.A. at Fisk University, and taught at all-black colleges Paul Quinn and Bishop College before returning to get her Ph.D. in mathematics from the University of Texas at Austin. She first applied to graduate school at Baylor University because it was in her home town, but she was denied admission because, as the director of admissions said, "We have not yet taken down the racial barrier here, although I have been hopeful that it would be done eventually. It seems that everyone is waiting for everyone else, and no one will take the initiative in such matters." Several years later, upon completing her doctorate, she was the first black professor hired at Baylor.

Marian Pour-El

Marian Pour-El received her Ph.D. in logic from Harvard University, though much of her doctoral work was done at Berkeley. She spent two years at the Institute of Advanced Study and worked with the famous mathematician Kurt Gödel. Now a full professor of mathematics at the University of Minnesota, she travels all over the world giving invited talks. Marian is one of the oldest women interviewed and yet in some ways has lived one of the most unconventional lives. She and her husband have explored all facets of long-distance relationships, a timely topic in these days when more and more professional couples are facing the "two-body problem."

Judy Roitman

Many people assume that mathematicians discover their passion for mathematics very early in life, but Judy Roitman did not. She was initially interested in English and linguistics, and received a bachelor's degree in English literature from Sarah Lawrence. It was not until graduate school that she realized that mathematics was her real passion, so she pursued a Ph.D. in mathematics at the University of California, Berkeley. Her research is in set-theoretic topology, and she worked closely with Mary Ellen Rudin (another mathematician in this book) at the University of Wisconsin. Given their advisee-mentor type relationship, it is fascinating to compare their attitudes about mathematics, and women in mathematics. Whereas Mary Ellen was relatively isolated as a woman in mathematics and is hesitant to talk about the topic, Judy has been extremely active in this area. Judy, like Lenore, was a president of the Association for Women in Mathematics, and has been involved in that organization from its birth. To what extent does identification with the label of "woman mathematician" help or hinder one's career? Judy is currently a full professor of mathematics at the University of Kansas. She is also a leader in the local Zen community and a published poet.

Mary Ellen Rudin

Mary Ellen Rudin is one of the leaders in the field of set-theoretic topology, having published more than seventy research articles. She has given invited addresses all over the world, and has served as vice-president of the American Mathematical Society. Despite her prominence in the mathematical community, she remained a lecturer at the University of Wisconsin for more than twenty years because of anti-nepotism rules. (Her husband is also a prominent mathematician at the University of Wisconsin.) Eventually the university was sufficiently embarrassed that it promoted her immediately to full professor. Mary Ellen has managed to be extremely active professionally while raising four children. Would she have been able to do that if she had had the full responsibilities of a tenured position?

Karen Uhlenbeck

Many mathematicians describe Karen as the best living woman research mathematician in the United States. A MacArthur Award winner, and member of the National Academy of Sciences and the American Academy of Arts and Sciences, she has won numerous fellowships and been a member of the Institute for Advanced Study at Princeton. A former vice-president of the American Mathematical Society, she is now a Sid Richardson Foundation Regents' Chair at the University of Texas at Austin. Despite these successes, Uhlenbeck still thinks of herself in some ways as an outsider. Her story illustrates that talent alone does not eliminate the problems women encounter.

When I began this project, the task seemed simple: to convey the stories of women in mathematics. I was near the end of my doctoral work in mathematics and had met very few women mathematicians. I was filled with questions: Were there women in mathematics? If so, what were their lives like? What was their relationship to their work? What drew them to mathematics, and what kept them there? Did they have children? What was it like for them to be women in this traditionally male field? How did they balance their personal and professional lives? Did they have advice for young women entering the field?

Since almost no literature existed on living women mathematicians, I began to do the research for this book. My working title captured this initial vision: "Dialogues with Women in Mathematics: Their Lives and Their Work." I interviewed a dozen prominent women around the country who represented a variety of mathematical fields, personal and professional situations, ages, geographic locations, and ethnic and racial backgrounds. Their stories were engaging, and became the basis of one part of this book.

My primary goal, initially, was to encourage women considering going into mathematics, as well as those already in mathematics, by breaking down prevalent stereotypes: either that women don't do math, or that those who do are of a particular (and often assumed to be unappealing) type. These real women's lives give vivid details of the diversity of experiences for women in math. They help replace stereotypes with a different ideal: an image of a multiplicity of models, so that other women feel that there is sufficient room to shape their own unique path in mathematics.

I would have liked to end the book here—a positive approach to women's underrepresentation in mathematics. The message: Since all of these women have created a niche for themselves in mathematics, other women can do the same.

But the deeper I got into this material, the more I came to realize that this approach to the book was only half the work. Two interrelated questions continued to surface, and were not adequately addressed by this initial vision. First, why is it that women continue to be significantly underrepresented in mathematics, particularly at the highest levels of accomplishment? And second, why is it that even the most successful women in mathematics, those who have already made it by standard measures of success, often continue to feel (to varying degrees) like outsiders in the mathematics community?

These issues could not simply be explained away by inequity in early schooling, or by stereotypes of women in society at large. Nor could they be reduced to the argument that women lack mathematical talent. For even women who had managed to escape the traditional expectations of women's roles, who clearly had a passion and a talent for mathematics, who managed to find their way into graduate programs, jobs, tenure, grants, and awards—even these women continued to face various kinds of resistance to women's presence.

It was clear, therefore, that it was not sufficient simply to supply more role models. Equally important was to look at the mathematics community itself. Are there ways in which the practices and ideology of this community create an atmosphere that prevents women from being completely accepted as full-fledged members?

A variety of scholars have begun to document the particular practices that create a "chilly climate" for women both in academia in general and in mathematics in particular.[1] And the women's stories in this book contribute to a deeper understanding of how these oftentimes unconscious behaviors can have a negative impact on women.

This book, however, goes beyond particular practices by trying to understand the deeper ideology that feeds them. What are the underlying assumptions and expectations of the mathematics community about who a mathematician is and what mathematics is all about? And how do these intersect with assumptions and expectations about women? Is there a tension between idealized images of a mathematician and images of women? Are there ways in which the mathematics community itself perpetuates the belief that women and mathematics are incompatible?

If these issues are not directly addressed, all of the women in this book can be written off as exceptions. If they are seen as *atypical* women, they do little to change attitudes about "typical" women. And indeed, as Margaret Rossiter points out in her book *Women Scientists in America*, this pattern has occurred repeatedly throughout history. Women's presence alone, particularly when they are a small minority, does not necessarily change deep-seated attitudes about women's place (or lack of place) in science. In addition, this kind of analysis is critical in order to understand why even those women who "made it," who were successful even within the standards of the mathematics community, often continued to feel like outsiders.

Thus, two central goals of the book became (1) to describe central components of the ideology of the mathematics community, and (2) to look at the impact of this ideology on women. By ideology I mean the underlying beliefs, attitudes, assumptions, and expectations of the community.

But identifying this ideology is a challenging task. Many of these assumptions and expectations are held at a preconscious level and are therefore often invisible both to those who benefit from them and to those who do not. Moreover, tradition has a powerful momentum. When one has been trained and spent many years invested in a given framework, it is hard to imagine that things could be different. Thus, tradition gets confused with necessity. "What has been" becomes "what must be."

How, then, does one identify this often invisible ideology? One strategy is to listen carefully to the stories of non-traditional members of the mathematics community—in this case women mathematicians. Since what we take most for granted is often least visible to us, by listening to people for whom traditional ways do not always work, we can begin to see more clearly what assumptions are embedded in those traditions.

Indeed, through the focus on women's lives, two sets of expectations about women become more clear: those held by society at large, and those held by the mathematics community in particular. To a large extent, the women I interviewed were able to break free of general society's stereotypical expectations of women, and create for themselves an image of who they were as women and as mathematicians. But it was harder to escape the expectations and stereotypes within the mathematics community— for this was their *chosen* community, and mathematics was such a significant part of their lives. In this sense, while they could use mathematics as a refuge from the confining expectations of society, it was harder to escape the sometimes confining expectations of mathematics.

A second strategy that I use to reveal ideology involves highlighting *imagery* that is prevalent in the mathematics community both about mathematicians and about mathematics. This imagery is important because it reveals underlying beliefs, assumptions, and expectations. Moreover, imagery not only reflects but affects who goes into mathematics, and how mathematics is practiced; in this way imagery can also *perpetuate* tradition.

Indeed, imagery can become a kind of gatekeeper, a way of defining who is an insider and who is an outsider. Those who conform to the traits embodied in the image of an archetypal mathematician are allowed in, while those who diverge too significantly from this norm are kept out. Often these judgments are made at an unconscious level. One develops an "instinct" about who will fit in as a mathematician, which is part of the learned wisdom of who is likely to succeed.

On an even more subtle level, one that is central to this book, ideology not only serves as a gatekeeper for who gets in, it also defines the tacit rules for those who do make it inside the castle walls. It defines the culture inside the community and, as I will argue, perpetuates an atmosphere that contributes to women's feeling marginalized.

I convey this imagery by quoting prominent mathematicians who have written about mathematics in the form of either essays or autobiographies. Because the mathematicians quoted tend to be highly respected individuals, their ideas often have a significant influence on the beliefs of the mathematics community.

Although facing some of these problems can, at times, be painful, I believe that in the end it is better for women, mathematics, and society at large to engage these issues than to pretend they don't exist. Only by bringing these issues forward can we recognize patterns. Individual behaviors, beliefs, and assumptions can seem small and insignificant, but when repeated patterns emerge that systematically discourage women, it is critical to articulate them in order to engage in constructive dialogue that can lead to real and lasting change.

At the same time, it is important to remember that the women in this book, and many others, have chosen to pursue mathematics because they *love* it. They delight in mathematics itself—they find it fun, exciting, challenging, and beautiful. For some it is like a wild and unpredictable adventure, for others it is a playful and pleasurable game, and for still others it is a delightfully beautiful art form. Indeed, most would agree with the sentiment of the mathematician Rozsa Peter, who wrote: "I love mathematics not only for its technical applications, but principally because it is beautiful; because man has breathed his spirit of play into it, and because it has given him his greatest game—the encompassing of the infinite."[2]

Organization of the Book

The book is organized into six chapters. Each challenges a dominant myth about who a mathematician is supposed to be or what mathematics is all about. Portraits of individual women accompany most of the chapters.

Depending on the interests and background of the reader, this book can be read in different ways. Some readers may be most interested in the lives of specific individuals; therefore, the biographical essays that accompany the chapters stand alone. They are a window into the lives of real women. While the majority of the profiles are written in an essay form, two are presented as interviews—primarily because the material insisted on it. The essays made it easier to highlight central issues and provide background information, while the interviews were better able to preserve the voice of the women themselves.[3]

Other readers may be more specifically interested in the central ideas that frame the book. Therefore, each chapter identifies underlying beliefs and assumptions of the mathematics community, the roots of those beliefs, and their impact on women. There is inevitably some overlap between the chapters and the biographies, since I use the words of these women to illustrate central points; however, the two parts of the book complement each other by examining major issues from different points of view.

Chapter 1 examines the myth that mathematicians work in complete isolation. Are mathematicians indeed the loners that they are often portrayed to be? I show that while a high degree of independence is in certain ways an asset—indeed, it is often used as a filter in selecting future mathematicians—in reality, mathematics is in many ways a highly social activity. Yet this social dimension often remains invisible. Moreover, women have more difficulty integrating into that social fabric, a fact that can hinder their development and recognition.

In chapter 2, I explore why it is so difficult to imagine a woman mathematician. The problem is not simply that few role models exist. I argue that there are underlying tensions between expectations of women and expectations of mathematicians. While other chapters examine this issue by focusing on images of mathematicians, this chapter highlights images of women. In particular, women are traditionally identified with specific roles, such as mothers or romantic partners, and these roles are typically identified with sets of expectations that are seen as incompatible with the life and attributes of a mathematician. I also explore strategies women have used to navigate through these conflicting expectations.

In chapter 3 I examine the image of mathematics as a "young man's game," and the deep-seated assumption that, like athletes, mathematicians are essentially useless beyond the age of thirty. This myth can have a significant impact on women who choose to have children. While the reality of these women's lives demonstrates that, for the most part, having children did *not* hinder their mathematical development, they still had to confront the powerful belief among the mathematics community that it would.[4]

Chapters 4 and 5 challenge two interrelated myths: that mathematics and politics don't mix, and that only white males do mathematics. Each of these chapters is accompanied by two biographical essays which flesh out the themes of the chapters in more depth.

While chapters 1 through 5 focus primarily on beliefs about mathematicians, chapter 6 focuses on beliefs about mathematics—though the two are invariably linked. In chapter 6 I argue that the image of a mathematician as a loner dovetails with the image of mathematics as a body of knowledge that is objective, true, and free of social influence. There is a strong investment in the image of mathematics as apolitical and value-free. It is what gives mathematics a distinctive kind of prestige. Since mathematical knowledge is seen as certain, absolute, and eternal, it is beyond doubt, transcending social or political bias—an ideal that most other disciplines strive to emulate. This vision of mathematics leaves little room for the possibility of bias with respect to gender.

However, mathematics is created by mathematicians. The social codes that shape mathematicians' lives have a significant bearing on the mathematics that is produced. Thus, an exploration of the ideology about *mathematicians* inevitably leads to an exploration of the ideology about *mathematics*. This is, without a doubt, the most challenging task of all. In chapter 6 I begin to lay out a framework which may be useful for further examination of this question. Ultimately, I believe it is impossible to extricate mathematics from its human and social context. Even at the very foundations of mathematics are values that are subject to both individual preference and social norms: in determining what counts as beautiful or important mathematics, even in determining what constitutes a proof. It is here that bias can inadvertently seep in. Exploring this subjective dimension of mathematics is critical for a deeper understanding of how gender intersects with mathematics.

Do Women Do Math Differently?

Having briefly described what this book is about, I should note what this book is *not* about. A question that often gets asked in reference to this work is, "Do women do mathematics differently?"

What a hornet's nest this question is. Most women in mathematics would be aghast at such an idea. They are as wedded as most male mathematicians to the idea that there is only one mathematics. Many would concur that whether a mathematician is male, female, black, white, or a Martian is irrelevant to the final products: theorems, proofs, and mathematical ideas. Thus, any affirmative answer to the question "Do women do mathematics differently?" would quickly alienate most mathematicians.

Furthermore, the question of whether women do mathematics differently can be interpreted in many ways. Do women ask different questions? Do they use different methods? Do they draw different conclusions? At this point it is difficult to determine answers to these questions. Women are still relatively few among established mathematicians, and if they diverged too strongly from the norm, they would not survive in the mathematical community.

On the other hand, there are issues that arise in looking at women's lives that need to be addressed. Women are more likely than men to feel marginalized; they are more likely to have lower levels of confidence, and are more likely to encounter subtle or overt obstacles to their development, acceptance, and recognition.

Thus, ultimately, the question "Do women do mathematics differently?" is not particularly useful. However, variations of that question are relevant. It *is* useful to ask whether women have different experiences in math, and whether they have a different relationship to their work. These questions are explored in the biographical essays. Furthermore, in the last chapter I ask not "Do women do mathematics differently?" but "Is it possible to think about mathematics differently?" and "What impact might that have on women?" This book, therefore, is not about trying to create or define something called "feminist mathematics." It is, instead, about trying to understand women's experiences in mathematics, and how those experiences are tied to the culture and beliefs of the mathematics community.

Two additional notes of caution are worth emphasizing at this point. In any discussion of "women's experience" or "impact on women," one must be very careful. While generalities can sometimes be useful, they do not capture *all* women's experience. In looking for patterns, one invariably loses some of the particularities. That is why this book includes the stories of eleven unique women and highlights their individuality.

Second, statements about women often get interpreted as being *only* about women. That is not the intent of this book. There is inevitably a large degree of overlap between male and female experiences in mathematics. It is useful to think of women's and men's experiences as overlapping bell curves. Because there is a great deal of variety within each curve, nothing definitive can be said about all women or all men. And because there is a great deal of overlap between the two curves, little can be said about only one group. But it *is* meaningful, and useful, to look for patterns. These patterns can reveal, for example, that the average experience for men might well be different from that for women. Moreover, these differences may suggest directions for further investigation. Finally, focusing on women's experiences can yield new perspectives on old problems or raise new questions as well as solutions.

However, identifying these differences is only half of the work. It is what we make of these differences that becomes the central concern.

There is a long history of women's "difference" being used against them, a way to justify restricting women to prescribed roles or creating rigid stereotypes. In this work, however, this information is used as a tool to help elucidate the culture of mathematics, and the ways in which that culture perpetuates those differences. It helps to identify what needs to change in the mathematics community so that women feel like equal members.

As long as women are being prevented or discouraged from pursuing mathematics simply because of their gender, that, to me, is unacceptable. But sometimes habits, customs, and assumptions are so ingrained that it is hard to recognize the ways in which gender bias is present. Looking at these women's lives is a way of shining a light on what has been hidden from view for a long time. It helps reveal how even the most "objective" of disciplines may not be immune from social bias. It also helps us understand how gender influences who becomes a mathematician, how it influences mathematicians' experience, and indirectly how it influences mathematics itself. Furthermore, only in making the invisible barriers visible can one hope to create real and lasting change. There is no reason mathematics cannot be inclusive of women and underrepresented groups. To make it so would ultimately benefit both men and women, as well as mathematics itself.

Historical Background and the Talent Question

Throughout history there has been a recurrent belief that at some fundamental level women were just no good at mathematics. First it was argued that their brains were too small, later that it would compromise their reproductive capacities, still later that their hormones were not compatible with mathematical development. These arguments were buttressed by the underlying belief that mathematics is ultimately a pure meritocracy. Those who have the gift would shine no matter what their background, sex, or race. As a corollary, it was assumed that if women were not excelling in the mathematical realm, they must simply lack the talent to compete. This belief in mathematics as a meritocracy is convenient; it absolves the mathematics community from any responsibility to change, and reduces the underrepresentation of women to a "women's problem." But a closer look at the history of women in mathematics reveals that in fact mathematics has been far from a pure meritocracy.

For centuries, if not millennia, women have faced myriad obstacles to becoming mathematicians. Up until the last century, women rarely had access to formal schooling, and had even less access to mathematical ideas. Many were able to learn mathematics only with the help of a male family member—some tagged along hungrily devouring scraps from the tutors hired for their brothers; others learned mathematics from their fathers. Most were banned from formal entrance to universities and devised a variety of strategies to overcome these obstacles. Sofia Kovalevskaia, for example, in the mid-nineteenth century, was not allowed to study in her native country of Russia, so she entered into a fictitious marriage in order to travel abroad and study unofficially with the famous mathematician Karl Weierstrass in Germany (unmarried women were not allowed to travel alone). Even then, since the German university would not allow Kovalevskaia to attend classes, Weierstrass had to give her private tutoring in his home. In the end, she wrote not one but *three* doctoral dissertations in order to be awarded her Ph.D.![5]

In France, the story was no different. At the end of the eighteenth century, Sophie Germain was not allowed to study in the newly formed École Polytechnique, which was established to train mathematicians and scientists for the country, simply because she was a woman. Through tenacious determination, however, she managed to get lecture notes from classes at the university and educate herself. She went on to win prestigious prizes for her work in mathematical physics.[6]

In the late nineteenth century, Grace Chisolm Young faced similar obstacles in her native England. Following in the footsteps of Kovalevskaia, she went to Germany to study mathematics, receiving the first official degree awarded to a woman in any subject whatsoever.[7] Ironically, Germany's generous spirit with respect to women in mathematics did not extend to its own citizens. German women at this time were not allowed entry into the universities, for that would lead to the undesirable problem of having to find them jobs after they received their degrees. But foreign women were, in this regard, irrelevant.

If the formal barriers were not enough, there were also strong social injunctions against women's pursuit of intellectual matters. Edward C. Clarke in his famous treatise *Sex and Education* (1873) powerfully and authoritatively admonished that such pursuits would endanger women's health. He argued that an overindulgence in matters of the mind would shrivel women's reproductive organs.

In a similar vein, the British mathematician Augustus De Morgan wrote about the budding mathematician Lady Byron Lovelace:

> I have never expressed to Lady Lovelace my opinion of her as a student of these matters. I always feared that it might promote an application to them which might be injurious to a person whose bodily health is not

strong. . . . But I feel bound to tell you that the power of thinking on these matters which Lady L. has always shown from the beginning of my correspondence with her, has been something so utterly out of the common way for any beginner, man or woman, but this power must be duly considered by her friends, with reference to the question whether they should urge or check her obvious determination to try not only to reach, but to go beyond, the present bounds of knowledge. . . .

All women who have published mathematics hitherto have shown knowledge, and power of getting it, but no one, except perhaps (I speak doubtfully) Maria Agnesi, has wrestled with difficulties and shows a man's strength in getting over them. The reason is obvious: the very great tension of mind which they require is beyond the strength of a woman's physical power of application.[8]

Thus, even if women gained access to a mentor and exhibited great talent, they were not necessarily encouraged to develop that talent. Indeed, they were more likely to be actively discouraged, often in the guise of concern for their own good.

Finally, even if women could navigate through the social and logistical pressures that prevented them from pursuing mathematics, it was almost impossible to find the jobs or receive the kind of community support and recognition that is so important to mathematical development. Getting a degree did not ensure acceptance or success. Women continued to be discriminated against in employment, membership in professional societies, recognition, and awards. Even Nobel Laureate winners were not admitted to the French Academy of Sciences if they were women. The oldest permanent scientific academy, the Royal Society of London, was founded in 1660. It did not admit its first woman until 1945. The French Academy of Sciences, founded in 1666, admitted its first woman, Yvonne Choquet-Bruhat, in 1979. Given these kinds of obstacles, it is hardly surprising that few women were able to survive and be productive in mathematics.[9]

The difficulty of assessing women's talent is further complicated by their invisibility. Often women did not use their real names in publications, or in correspondence with male colleagues, for fear of not being taken seriously. Sophie Germain, for example, used the pen name "Monsieur LeBlanc" in corresponding with the famous mathematicians Carl Friedrich Gauss and Joseph Louis Lagrange. Lagrange was so impressed by her work that he insisted on meeting "him" and was astounded to discover that Monsieur LeBlanc was in fact a woman![10] Other women used initials instead of false names, leading people to assume, again, that they were male.

Moreover, many women did joint work with a husband or male colleague. This was often the only way they could gain access to equipment or research materials. But typically it was the man who was remembered and cited for the work. In more extreme cases, the woman's name was not published at all. This was true of the mathematician Grace Chisolm Young, who did a great deal of joint work with her husband, William Young. Some of this work was published under both of their names (e.g., a set theory textbook), but most was published only under his name. Her husband justified this practice in a letter he wrote to her:

> The fact is that our papers ought to be published under our joint names, but if this were done neither of us [would] get the benefit of it. No. Mine the laurels now and the knowledge. Yours the knowledge only. Everything under my name now, and later when the loaves and fishes are no more procurable in that way, everything or much under your name.
>
> There is my programme. At present you can't undertake a public career. You have your children. I can and do.[11]

Furthermore, even when women did publish work under their own names, they often were forgotten. A notable example is the physicist Lise Meitner, one of the discoverers of nuclear fission, who even today is often not mentioned in histories of the subject.[12]

The difficulty in assessing women's mathematical talent is further exacerbated by the fact that even when women did carve out an important niche for themselves, their work was often later trivialized or devalued. Some of the notable women in mathematics in the eighteenth and nineteenth centuries made important contributions by synthesizing and popularizing the mathematics that was being done, but which few people had access to. For example, the calculus textbook written by Maria Agnesi was the most important text of the period, for it brought together the work of mathematicians who had virtually no communication with each other, yet whose work, although not their styles, had much in common. Thus her work was particularly useful in facilitating communication and development of calculus.

The kinds of obstacles women faced evolved over time. Through a variety of intriguing strategies, by the late 1800s women were gaining access to college education, and by the early 1900s they had largely dismantled the significant barriers to graduate education. With the critical badge of formal membership that a Ph.D. provided, women's complete acceptance and integration into mathematics seemed imminent. But as Margaret Rossiter documents in her book *Women Scientists in America*, complete integration as professionals continued to be an elusive dream.

The ante continued to rise; no longer was a college education, or even a doctorate, sufficient credentials for membership in the mathematical elite. Membership and recognition in the professional societies became increasingly important. It was this inner circle that played a powerful role in shaping future mathematics as well as our vision of what it means to be a mathematician, yet women were often formally or informally excluded from this inner circle.

Nonetheless, throughout the first half of this century, women did make significant inroads. They were able to get jobs as mathematics professors, at least at women's colleges. And while the high teaching loads at these newly founded, and hence poorly endowed, colleges precluded extensive commitments to research, these pioneering women professors did continue to be mathematically active, and encouraged many younger women to pursue mathematics as well.

Recent Past

Progress for women in mathematics, even in the last century, was not necessarily linear. It was constantly subject to the fits and starts of social change. For example, during the conservative postwar period of the 1950s and 1960s, when the social pressure on women to stay home and allow men to obtain employment was at its peak, women's participation in mathematics was at an all-time low for the century.[13] Only later, with the fire of the Sputnik era and the subsequent women's movement, did women regain and begin to significantly increase their numbers in mathematics and science.

By the 1970s, the nature of the discussion had changed. No longer was the focus simply on gaining *access* to mathematics; the debate had shifted in a significant way, and the central question was now how to attain *equity*. This change in language reveals a shift in underlying assumptions about women's place in society. A fight for access does not necessarily challenge deep-seated expectations about women's roles, for the few who pursue these unconventional paths can be seen as exceptions. But with the focus on equity rather than access, the assumption was that, at a fundamental level, women were no different from men in their ability to do mathematics; there was no reason why a woman shouldn't go into mathematics. This was a very different attitude about women's place in society.

Many scholars and educators, however, were concerned and puzzled by the fact that even as formal barriers were disappearing, women were still not pursuing mathematics to the extent that men were. Why was it that even as equal opportunity was being achieved, equal outcome was not? Why weren't more women choosing to study math? The professional and financial implications for their choices were significant. As Lucy Sells described it, mathematics serves as a critical filter for future employment opportunities.[14] By not pursuing math, women were locking themselves out of the more prestigious and lucrative careers in science, medicine, and technology.

A number of studies were commissioned to examine this phenomenon. The simplistic assumption that women didn't do mathematics because they lacked talent was replaced by a far more sophisticated analysis of the many variables that contribute to women's underrepresentation. In the 1970s the National Institute of Education sponsored a series of studies specifically to address the question of women's lower participation in mathematics.[15] Most of this work focused on the pre-college level. At its core was an attempt to understand how to change women, the educational environment, and society at large, so that women's pursuit of mathematics seemed natural and was encouraged.[16]

A variety of strategies emerged from this period, but perhaps one of the most effective was making four years of high school mathematics mandatory. This prevented women from prematurely locking themselves out of future career paths. With this stronger high school background, they now at least had the option of pursuing college mathematics, which was, and is, a prerequisite to most careers in science, medicine, and technology.

Indeed, the efforts to increase women's participation in mathematics during the past two decades have been largely successful. Most striking is the fact that women now constitute approximately 44 percent of the mathematics majors in the country. It was once assumed that women did not have the biological makeup to do trigonometry and algebra, yet they are now studying differential equations and topology. Thus, as history demonstrates, "insufficient talent" is an insufficient explanation for women's underrepresentation in mathematics. Institutions and society can have a dramatic effect on who pursues mathematics.

It would be tempting to believe that we have finally reached a happy conclusion, and our story can end here. Unfortunately, the full story is not so simple. As several recent studies remind us, we should be cautious in our optimism, for we are still quite a way from true equity.

These studies reveal that inequity continues to be present in the attitudes, expectations, treatment, and outcome of students. In classrooms, women speak less frequently and for shorter amounts of time than male students, and they receive less attention from, and are less challenged by, their teachers.[17] Men are more likely to be encouraged to develop independent learning styles, which are important in future cognitive development.[18] Women are more averse to the competitive style that characterizes most mathematics and science classrooms.[19] Women are less likely to be encouraged to pursue mathematics and science careers. Furthermore, women are more likely to attribute success to luck and failure to ability, while men attribute success to ability and failure to lack of effort. This difference means that failure is more likely to lead women to drop out of mathematics, and contributes to their decline in self-esteem and confidence, while men are more likely to conclude that they just need to work harder and thus don't question their ability to the same degree.[20]

Even encouraging statistics—such as that women constitute 44 percent of the nation's mathematics majors—can mask the reality that many women face. First, women still do not necessarily experience this parity in many of their classes, particularly in the first two of years of college, since many physics, engineering, and other hard science majors take first- and second-year mathematics courses, and these fields have a much lower percentage of women. Second, the almost equal percentage of women majors is in part a reflection of the fact that fewer men are majoring in math, and therefore the total number of mathematics majors is significantly lower than it was twenty years ago.[21] Finally, the percentage of women majoring in mathematics at the top (Group 1) institutions in the country is still significantly lower, approximately 37 percent.[22]

Moreover, beyond the bachelor's level, women's participation in mathematics drops significantly. Women make up only 24 percent of the Ph.D.'s in mathematics.[23] In the most prestigious Group 1 institutions (39 of the top research-doctorate programs in the country), women made up only 17.3 percent of 1991–92 doctorates.[24] Yet these top institutions produce the lion's share of future top researchers, high-profile mathematicians, and mathematicians with the most political power.

With respect to employment figures, the percentages of women decrease even further. Women are only 19 percent of full-time faculty, and less than 10 percent of tenured professors in mathematics. If we look at doctoral granting departments, the percentage of tenured women drops to less than 5 percent.[25]

That women are far from achieving equity in terms of outcome is inescapable. The number of women in the uppermost echelons of "success" is infinitesimal. No woman has won a Field's Medal (often considered the Nobel Prize in mathematics, since the Nobel Prize is not awarded in mathematics), and only four women mathematicians have been admitted to the National Academy of Sciences. No women before 1990 were awarded prizes by the American Mathematical Society.

It would be easy to return to the talent question. Perhaps women are just not as gifted as their male counterparts. But there is no reason to believe that this assumption is any more sufficient or justified now than it was historically. In Ravenna Helson's study of some of the most creative mathematicians in the country, she found no difference between the men and women in this category with respect to the age at which they received their Ph.D., or the age at which their first paper was published. But later, the men published more papers and held more important positions at prestigious institutions. Only two or three of the creative women taught graduate students, and one-third, including some of the highest rated, had no regular position at all.[26] Men who are this gifted rarely face such difficulties in obtaining appropriate employment.

Other studies, as well as the interviews for this book, show clearly that obstacles continue to exist to women's complete acceptance in mathematics. Though these obstacles are rarely the blatant or formal barriers of the past, they continue to exist in more subtle forms, embedded in attitudes, beliefs, and expectations about women, mathematicians, and mathematics.

Since women are even further from equity the farther along in the pipeline we go, the focus of this book is on the graduate and professional level; this is where the percentages of women are most discouraging, and where the least work has been done.

The way in which one examines equity at these stages must necessarily be different. We are dealing with a preselected group of people—women who are highly motivated to pursue mathematics—but a group who, despite their passion for math, often still feel marginalized. Thus, while much of the earlier work on women in mathematics focused on how to turn women on to mathematics, the primary focus of this book is on how to prevent them from being turned off, though the two are inevitably connected.

I examine this question by asking not just how might women change to adapt to the world of mathematics, but how might the mathematics community change to more fully embrace women?

WOMEN IN MATHEMATICS

1

Rugged Individualism and the Mathematical Marlboro Man

Myth: Mathematicians work in complete isolation.

During high school, I would often get calls late in the evening from one or another of my classmates saying that she was stuck on a certain problem on the math homework and was wondering whether I had been able to solve it. If I hadn't started it, we would thrash it out together on the phone, experimenting with different ideas until we reached an answer. If it was too hard to do on the phone, sometimes we would hang up, work for a while on our own, and reconnect later. It was a fun, cooperative, and productive interchange. We were all girls, at an all-girls' high school—we enjoyed math, and it never occurred to us that mathematics and being a woman were incompatible, or that doing math was socially isolating.

When I got to college, I experienced the other extreme. I was in one of the most advanced courses a freshman could be placed in, at one of the top universities in the country. And I was one of the only women. My classmates never seemed to talk with each other, or at least no one talked with me. The classes were, in general, completely impersonal. With the exception of two visiting professors from other institutions, teaching was a low priority, and there was no personal contact or encouragement from professors. So I learned to work completely alone. I didn't mind. I liked working alone—the freedom to study in my own way, pursue my own ideas, and claim complete credit for insights and accomplishments. But over the years I felt myself growing distant from mathematics. I still liked math itself, but it seemed bereft of human contact: friendships, play, and the sharing of exciting ideas. Mathematics was still so much a part of my identity that I was not willing to give it up, but I found myself living a double life. There was my inner mental world of mathematics, which was a beautiful and fascinating puzzle that I worked on alone, and then there were other outlets for my social needs. I double majored in philosophy, through which I developed friendships with faculty and graduate students. They gave me a sense of what it is like to live an academic life and encouraged me to apply for fellowships and graduate school. In other spheres such as volleyball, I learned about group dynamics: how a whole can be more than the sum of its parts; how important support and encouragement are to growth; how physical, mental, and emotional growth are intimately intertwined. But my mathematics world never mixed with my social world.

The pattern of working alone on mathematics was so deeply ingrained throughout college that by the time I started graduate school, I no longer knew how to do mathematics with others. And I lacked the confidence that was required to relearn how to do it—for it meant being willing to make mistakes and reveal deficiencies in front of peers. Graduate school is usually a time of proving oneself (often at the expense of others); it is not an easy time to take risks or acknowledge weakness. So I continued my inward trajectory.

I thought that the change I had experienced in going from high school to college was inevitable—and that if I was going to continue in mathematics, I would have to learn to enjoy it alone. As Freeman Dyson, the famous physicist who began his intellectual career as a mathematician, writes, "There are always exceptions to every rule, but statistically I have found the physicists to be more gregarious, and they on the whole have a more communal enterprise going which I enjoy. . . . Most of the mathematicians I knew were rather lonely people. . . . The mathematicians tended to be each working alone in a corner."[1] Over time, however, I began to question this belief: Are mathematicians in fact so isolated? And if they are, is this indeed necessary?

My interviews with women mathematicians across the country initially seemed to yield contradictory answers to these questions. On one hand, almost all of the women were highly independent and autonomous individuals; it was a significant part of their identity. On the other hand, many of the women also talked about how important ties to the mathematics community were in their development and success. These ties operate on many levels: they open doors, provide encouragement and support, teach the methods, language, and implicit rules of the discipline, supply ideas and problems, and facilitate recognition and awards. These ties not only influence what might be considered "external factors"—such as gaining entry to the world of mathematics—they often play a significant role in the actual *doing* of mathematics as well.

How, then, does one reconcile the highly individualistic nature of many mathematicians (both men and women) with the reality that social ties to the math community play such an important role in mathematicians' lives?

I would argue that the dominant image of a mathematician is that of a loner, and this image serves as a filter that influences who chooses to go into mathematics, and whom the mathematics community takes as one of its own. Those who make it through this filter do indeed tend to be highly autonomous individuals who are comfortable, may even prefer, working alone.

At the same time, at all stages, ties to the mathematics community profoundly shape one's development and success. It may be true that one can be a mathematician and be relatively isolated, but most successful mathematicians—both men and women—have had important connections that have had a profound impact on their research and professional life. In this way, the image of the isolated mathematician is both misleading and inaccurate.

Moreover, women, in general, have a more difficult time integrating into the social fabric of mathematics than do men. And yet the difficulties women face are masked by the belief in the rugged individualist. Because mathematicians are depicted as loners, their ties to the mathematics community not only become invisible, they are also assumed to be irrelevant. However, the truth is that those ties can make an enormous difference to productivity and success.

Images of Mathematicians

When most people think of a mathematician, they picture a nerd engrossed in scribbles and equations, a calculator in one hand, chalk in the other, oblivious to anything but numbers and geometric shapes—not an image most people would aspire to. The image of a mathematician within the mathematics community, however, is quite different. It is a romantic image of an explorer, living a life filled with adventure, discovery, and excitement. As one mathematician, Judy Roitman, says, it's like being a deep-sea diver, diving into an endless ocean, and being constantly delighted by the beautiful discoveries that lie below the surface.

But the image of the explorer often takes a particular twist that is allied with stereotypical images more of men than of women. Mathematicians are sometimes portrayed as intellectual cowboys out to tame the mathematical universe—what one might describe as a Mathematical Marlboro Man. Indeed, mathematics has been described as "the science which lassos the flying stars."[2] Mathematicians are depicted as living heroic lives, filled with self-sacrifice, all in the name of the search for truth.[3] Their lives are often portrayed as either a quest or an adventure that involves great risk: "Many times a scientific truth is placed as it were on a lofty peak, and to reach it we have at our disposal at first only hard paths along perilous slopes whence it is easy to fall into the abysses where dwells error; only after we have reached the peak by these paths is it possible to lay out safe roads which lead there without peril."[4]

In this context, the duty of a mathematician, like that of a cowboy, is to tame the unruly and unexplored mathematical landscape. But instead of trying to tame horses or cattle, mathematicians tame creatures such as infinity. The mathematician James Pierpont writes, "The notion of infinity is our greatest friend; it is also the greatest enemy of our peace of mind. . . . Weierstrass taught us to believe that we had at last thoroughly tamed and domesticated this unruly element. Such however is not the case; it has broken loose again and Hilbert and Brouwer have set out to tame it once more. For how long? We wonder."[5]

Even the primary tools that a mathematician uses—imagination and reason—are sometimes portrayed as something akin to a horse and spear. Imagination has been described as "the roadway through the forest in order that cold logic might follow." And reason gives "penetrating power to the mind which enables it to pierce a subject to its core and discover its elements."[6]

However, the image of the Mathematical Marlboro Man—a tough, independent hero taming uncharted terrain—is targeted to attracting men more than women. Though some women are drawn to the image of an intellectual cowboy, the associated traits are traditionally identified with men and masculinity. Hence, these images fuel the belief that mathematics is a male domain—an idea that is then absorbed both by the mathematics community and by the general public—and can indirectly discourage women from pursuing mathematics.

In earlier times, women were actively discouraged from pursuing mathematics because it was assumed that they did not have the physical strength to sustain such mental focus. The mathematician Augustus De Morgan was ambivalent about encouraging Lady Byron Lovelace to pursue mathematics, though her talent was beyond question, because he believed that doing mathematics required "a man's strength," and he worried that she, as a woman, would not be able to bear "the very great tension of mind" to which mathematics would lead her.[7] Mathematicians are assumed to be tough; women are not. Though that kind of direct assumption of incompatibility is far less common today, vestiges of it continue to exist.

RUGGED INDIVIDUALISM

One of the most prominent traits associated with the Mathematical Marlboro Man is a kind of rugged individualism—a belief that mathematicians don't need anyone, that they are primarily self-sufficient—indeed, that they work essentially alone.[8] As the mathematician Camille Jordan writes: "Continue in scientific research, you will experience great joy from it. But you must learn to enjoy it alone. You will be a subject of astonishment to those close to you. You will not be much better understood by the scholarly world. Mathematicians have a place apart there, and even they do not always read each other."[9]

Others have written that mathematicians must be comfortable with relative isolation for days, weeks, even months at a time. As Donald Weidman says in his article "Emotional Perils of Mathematics":

> The mathematician must be capable of total involvement in a specific problem. To do mathematics, you must immerse yourself completely in a situation, studying it from all aspects, toying with it day and night, and devoting every scrap of available energy to understanding it. You can permit yourself occasional breaks, and probably should; nevertheless the state of immersion must go on for somewhat extended periods, usually several days or weeks.[10]

This image of a mathematician as a loner is reinforced by media coverage. One of the most famous mathematicians in recent decades, Andrew Wiles, who solved what has been labeled "the world's most famous math problem"—Fermat's Last Theorem—is described as having worked in complete isolation for seven years "on his secret project sitting in a barren attic office on the third floor of his Tudor style house. No computer was necessary and no telephone was present to intrude on the absolute silence."[11]

On the surface, then, it would appear that most mathematicians are loners. And because this is the dominant image, to some extent it has the potential of self-fulfillment. Those who think of themselves as highly independent and are comfortable working alone are more likely to be drawn to this vision of mathematics, while those who crave work of a more social nature are turned off from a path that seems so isolated. Moreover, the mathematics community develops a template of who a mathematician is supposed to be, and decisions about who is accepted into the community are based on this template. One of the characteristics that have come to be seen as central is a high degree of independence.

This is just as true of women as it is of men. Indeed, women may need to be even more independent, because not only are they more likely to become ostracized from their general community by breaking with traditional roles and pursuing mathematics, they are also more likely to be isolated within the mathematics community because they are women.

WOMEN'S AUTONOMY

Not surprisingly, therefore, virtually all of the women interviewed for this book exhibited a high degree of autonomy and independence,[12] as well as a willingness and ability to work alone. Most developed these traits at an early age. The roots of that autonomy varied: for some it was a result of physical isolation, for others of social isolation. Some chose it, others did not. But almost all embraced it; being alone, or different, was a source of creativity.

In many of these cases it is hard to identify whether these women's independence was more a product of their personalities or a response to their external environment. Usually it was a blending of the two.

In the case of Fern Hunt, a mathematician formerly at Howard University, both her personality and being black in a largely white educational setting contributed to her independence. The social isolation she felt as early as elementary school continued in high school at Bronx Science, where she was a minority both as a woman and as an African American. She found some companionship from a variety of sources—a book club, her church, a Saturday science program—but for the most part she worked and learned alone. Her independence helped channel her feeling of isolation into creative pursuits.

For Mary Ellen Rudin, a mathematician at the University of Wisconsin, geographic isolation in early childhood influenced her mathematical development. When she was six, her family moved to a small town in southwest Texas.

> It is about 3000 feet high in the canyon formed by the Frio river with mountains on all sides. In those days you entered the town by going fifty miles up a dirt road—you had to ford the river seven times to get there. . . . It was a real mountain community. Many kids came to school on horseback. . . . Most people didn't have running water or any of the things you think of as being perfectly standard. . . .
>
> We also had a lot of time to develop games. We had few toys. There was no movie house in town. We listened to things on the radio. That was our

only contact with the outside world. But our games were very elaborate and purely in the imagination. I think actually that that is something that contributes to making a mathematician—having some time to think and being in the habit of imagining all sorts of complicated things.[13]

In the case of Karen Uhlenbeck, a distinguished mathematician at the University of Texas, autonomy developed in part as a rejection of the traditional expectations of women. Uhlenbeck says about her childhood,

> I really felt that I didn't get along with people very well. I always had a lot of girlfriends, but I never had very many boyfriends. I never felt really like I was part of anything. I went my own way, without really wanting to, but I never did understand the trick of doing things like you were supposed to. . . . I did a lot of escaping by being interested in stuff. Maybe by that stage I wasn't about to drop my interest in things in order to behave properly with people. . . . You need to remember this was the fifties. I did not feel like I was supposed to do anything interesting except date boys. That was what girls did.

One symptom, or source, of that independence which was almost universal in their stories was the degree to which they immersed themselves in books from a very early age. For Judy Roitman, a topologist at the University of Kansas, this was a result of a persistent cough that she had as a small child, which precluded her from participating in physical activities with her peers. She insisted, at the age of four, that her father teach her how to read. From then on, books became her closest companions. Fern Hunt also found companionship in books; she was stimulated by the MAA New Mathematics Library Series, books aimed at a high school audience, which many have found an alluring introduction to mathematics. Karen Uhlenbeck used books to discover an alternative vision of the world, reading not only every book in her library on mathematics and science, but as many books as she could find about frontier life as well.

Thus, while so many of these women experienced various forms of isolation in their early years—sometimes a happy isolation, other times not—almost all came to appreciate and embrace working and learning alone. For all of them it was a source of creative energy; it gave them the opportunity to find other things they loved. The intellectual independence and focus they had also served them well throughout their schooling and early professional life.

Ravenna Helson, in her study of creative women mathematicians, found a similar pattern. As she reports, all of the women that she studied valued their independence and autonomy. This autonomy was often part of a rejection of traditional expectations. "They [the creative women mathematicians] are not bound by, do not recognize, and perhaps somewhat stubbornly resist, conventional [societal] patterns."

Thus, the vision of mathematicians as rugged individualists does not turn all women off of mathematics. Many women have found their way in, created a niche for themselves, and love doing mathematics. At the same time, we are left with several questions: How many people *are* turned off of mathematics because of traditional images of mathematicians, and in particular the belief that mathematicians lead isolated lives? Second, are women more likely than men to be turned off by this vision?

More research is needed on both of these questions; almost no literature explicitly looks at the link between gender and the perception of mathematics as a highly individualistic pursuit. However, studies that have been done show that the images and beliefs women hold about mathematics (as well as images and beliefs about women) *can* play an important role in the path they choose. As Gilah Leder, a specialist in the area of gender and education, summarizes,

> When choices are available, the path selected is influenced not simply by reality but also by the individual's perceptions and interpretations of reality. Thus students capable of continuing with mathematics but who *believe* that the study of mathematics is inappropriate for them may select themselves out of mathematics or perform at a level they believe others consider appropriate for them. . . .
> Studies suggest that females' lower performance in mathematics is a function not so much of ability per se as of internalization of, and conforming to, the expectations of others.[14]

Therefore, because images play a significant role in shaping who goes into mathematics, it is important to ask whether the current images are indeed accurate and necessary.

Mathematics as a Social Activity

Is it true, then, that mathematicians are loners? Do they work in isolation? A closer look reveals that the private world of mathematics is not so clear-cut as this public image suggests. In fact, mathematicians define a whole spectrum of work styles; some work in almost complete isolation for fairly long periods of time, while others do the majority of their work in collaboration with other mathematicians. Most fall somewhere in between these two extremes—some are fairly social and interact with other mathematicians on a regular basis getting ideas, learning new material, being exposed to new problems, but then may take one of these problems and work on it in solitude, like a squirrel privately cracking and devouring a nut in its nest. Much depends on a person's work style, as well as the current stage of his or her career, or on a particular problem. Although the focus on rugged individualism can act as a filter for potential mathematicians, and a high degree of independence can be a useful trait in mathematics, the dominant image of a mathematician as a loner captures only one end of a spectrum of styles, and in this way is incomplete, misleading, and to a large degree inaccurate.

In fact, mathematics is in many ways an extremely social activity, and the mathematical community plays a powerful role in shaping both the lives and the work of its members. The isolation attributed to Andrew Wiles is the exception rather than the rule; moreover, even in Wiles's case, interaction with the mathematical community played an important role in several key ways. Yet, because the image of a mathematician as an independent, self-sufficient individual has such a powerful hold, even those inside the mathematics community often do not fully recognize the significant role that community plays in shaping a mathematician's life.

Community operates at all stages of a mathematician's development: initial motivation, establishing a mathematical identity, getting a job, continued development, promotion, and recognition. Though the examples presented here are drawn largely from the women interviewed for this book, community plays an important role in men's lives as well. But because women are surrounded by messages that mathematics is not a woman's domain, and because they are more likely to encounter obstacles to their presence in mathematics, community support is particularly important in their lives, and is therefore the primary focus in the discussion that follows.

INITIAL MOTIVATION

The myth and the reality of how one enters the world of mathematics are often at odds, particularly for women, who are less likely to view the pursuit of mathematics as a "natural path." The image of mathematicians as highly autonomous and self-reliant suggests that they should be completely motivated from within, driven simply by the love of mathematics, a passion that takes hold and does not let them go. In this image, dependence on external sources of motivation is seen as a weakness.

But in reality, while internal drive and a love of mathematics are necessary ingredients for a life of mathematics, they are often not sufficient. It is not uncommon for internal motivation to be triggered by another person, for example, by a teacher or family member.

All of the women interviewed did delight in the discovery of mathematics, and this love of mathematics was a major motivation in their pursuing it. Many described an experience similar to Judy Roitman's when she was introduced to the concept of infinity in junior high school: "The whole idea that you could work with infinite quantities and be precise about it just blew me away." However, that connection to mathematics was usually facilitated by a teacher who made the subject come to life, or who was particularly encouraging and affirming. This was even more true of the women of color, who were most likely to be raised with messages that they were not good enough or that mathematics was not a viable option for them. In the case of Fern Hunt, for example, it was a high school biology teacher (who was also her first black teacher) who first turned her on to a life of science. He challenged her, believed in her, and encouraged her to pursue paths she did not even know were open to her. For Vivienne Malone-Mayes, the first black faculty member hired at Baylor University, it was two professors at Fisk University who had a powerful influence on her decision to pursue mathematics. (One was Evelyn Boyd Granville, one of the first two black women to receive a Ph.D. in mathematics in America.) They helped her see the beauty and power of mathematics, and gave her a taste of what it meant to "make mathematics her own."

These stories reveal that being self-motivated does not preclude being stimulated and sustained by external forces, including parents, teachers, friends, and community. The role of other people in sustaining one's mathematical drive is even more dramatic as we move from the initial enticing experience to the later stages of one's mathematical development. This is where the myth of the lone cowboy really begins to break down.

Establishing a Mathematical Identity

Once the mathematical fire is lit, it must be sustained by a community of peers who help stoke it. A critical role of community is to create a sense of belonging, a kind of mathematical family with whom one identifies. For women this can be particularly important because women who are deeply interested in mathematics are more likely to be socially ostracized than men. Marcia Groszek, a mathematician at Dartmouth College, found this sense of community in the Hampshire College Summer Mathematics Program, which she attended during high school. She was accepted both as a woman and as someone interested in math.

> It put me in a situation where I wasn't automatically a complete oddball for being interested in math. I went through my entire junior high career

being a social outcast, just assuming that I was completely incompetent in that realm. At Hampshire I was thrown into this other situation, with other people who assumed that being in mathematics is okay, people who could look at me and who could see what I really look like.

What is most strikingly consistent in the lives of these women mathematicians is that almost all of them found such a community—particularly during graduate school—and that without that connection, it is unlikely they would have persisted or succeeded in mathematics to the degree that they did. Graduate school is the critical stage for making connections with future colleagues and collaborators. It is when students develop a sense of identity as mathematicians, and it is the first time they do serious independent mathematical research. Thus, a mathematician's self-image and relationship to the mathematics community are strongly influenced by this apprenticeship period during graduate school.

Fern Hunt, for example, developed an important sense of community not only with fellow graduate students at New York University, but also by joining a group of black mathematicians who served as a support network. Fan Chung, a researcher and manager at Bellcore (an offshoot of Bell Labs), went to college in Taiwan, where her experience was closer to our version of graduate school. It was there that she formed a close community of women friends and peers in mathematics, most of whom later came to the United States for both Ph.D.'s and jobs in mathematics. The camaraderie among this group was particularly important because they were not only foreigners in their home country as women pursuing mathematics, they were also foreigners to the United States. In both cases, having each other as inspiration and support helped them to deal with the challenges that inevitably arose.

Almost none of the women consciously thought in terms of creating a community; it was something that happened to them. This was most apparent in the case of Mary Ellen Rudin, who was never without community. From her first day of college at the University of Texas, when the famous mathematician R. L. Moore took her under his wing, she became part of a community of mathematicians, sometimes referred to as the Moore family, which provided mathematical stimulation and colleagues for the rest of her career.

This sense of community is important for men as well as women. However, certain pulls are more likely to arise for women that make community support particularly important, for example, when they have children. Linda Keen, a mathematician at City University of New York, described what it was like when she had her first child. Her neighbors felt she should be home full-time, and there was little encouragement in that social context for her to return to graduate work. It was the mathematics community at New York University that drew her back in.

> Part of me wanted to stay home and just be with the kid. But I also wanted to keep doing what I was doing. I certainly couldn't imagine always staying home, and I was afraid that if I stopped I would cut off those options. I liked the feedback of being accepted as a peer on a work level either in graduate school or afterward. It was very important for my own sense of self. . . . I was getting a lot of good feedback from being involved in mathematics and the work community and the social community that went with it. I was not ready to give that up.

Much of her identity and positive sense of self were tied to the mathematical community. Had that community been discouraging rather than encouraging to her, it would have been harder to maintain the determination, focus, and energy that it takes to balance children with mathematics.

Not all women (nor all men) are so fortunate to find such a supportive community in graduate school. As a result, some leave mathematics altogether, while others continue but are stifled in their development. Some respond by finding alternative communities. Judy Roitman, for example, was a graduate student at the University of California at Berkeley. Because it was a difficult place to be a woman in mathematics, she developed a secondary community at the University of Wisconsin, where she found a stimulating group of mathematicians in her field and a female mentor. She also became actively involved in the newly formed Association for Women in Mathematics.

Vivienne Malone-Mayes was one of the few women interviewed who had a distinctly negative experience at the beginning of graduate school. Being both black and female made her doubly isolated. As she recalled, "Although there were two other girls in the class, they avoided me like I was some sort of plague, because if you're not a white woman, you can't associate with anybody except maybe handicapped men." As time went on, she became more accepted in the community, but Malone-Mayes was never able to develop the same kind of professional support network for research that other women interviewed for this book did. It is likely that this had an impact on the number of research publications she later produced. There were certainly other factors that influenced the amount of research she was able to do; she had other priorities, other commitments, and other pressures. However, when she did receive support, for example in the form of a graduate scholarship from the American Association for University Women, it had a tremendous influence on her belief in herself, and her determination to pursue graduate work despite the obstacles she encountered. At a professional level, she might well have benefited from the kind of network that Fern Hunt found in New York among black mathematicians. Instead, Malone-Mayes faced almost complete isolation as both a woman and an African American in mathematics at Baylor University. That isolation took its toll.

FINDING A JOB

Neither initial access to mathematics nor early training accords with the image of a mathematician as a loner, particularly among women mathematicians. But what about once one is a certified mathematician?

Even here, we see that mathematics is in fact a very social activity. Almost all of the women interviewed said that forming connections with a critical person or group of people in their field was key to their success in mathematics. This could be an advisor, a mentor, a patron, or even a small group of powerful people. Such a person becomes one's advocate and plays a major role in establishing and promoting one in the mathematical community. The first time this comes into play is in finding a job.

It is often assumed that getting a job after graduate school is based purely on merit, but the reality is quite different. As Karen Uhlenbeck points out, connections are extremely important:

> That's how you get a job. It's really bad the last few years. There were no jobs for a while, so that wasn't good. It hasn't gotten any better in the abstract because every place like this gets 750 job applications. We can't process that. So you hire people that you hear about—which means that your pals call you up. So it's gone back to the "good old boy" system without any question because we can't handle the paperwork. Nobody really desires that. We would like to look at the applications and decide who is more suitable to the department. For instructors you want people you think are good combined with people who are suited to the research environment here. You don't pick necessarily the best people. You pick the people that you think will fit in and will benefit. [But because of the paperwork], you hire people that you know. For instructors it's pretty much who your friends are out there, because they haven't even had the opportunity to publish their work.

Advisors play a critical role at this stage; they act as mediators between the recent graduate and the mathematical community by promoting their students and helping them gain public recognition.[15] Therefore, *who* one's advisor is, and how well connected he or she is to the mathematical community, are quite important.[16] Many interviewees stated that when young mathematicians were looking for a job, their advisor would get on the phone and call friends and colleagues. The assumption was that the favor would be exchanged in the future. It was considered shocking and embarrassing to have to actually go out and apply for jobs. That's not what strong candidates did.

Mary Ellen Rudin, for example, has never applied for a job in her life. R. L. Moore, her advisor for both undergraduate and graduate school, made all the arrangements for her first position at Duke without even consulting her. She knew nothing about it until she was informed that that is where she was expected to go. Another colleague later arranged for her to receive a fellowship for which she had never even applied. These behind-the-scenes informal arrangements were standard for students coming out of research institutions.

The informal nature of these avenues for finding jobs must be stressed. It is the "softness" of this process that can make it hard to pinpoint the ways that subtle bias can be built into the system. When recommendations are made by word of mouth, or unsolicited comments conveyed, there is often no paper trail to document inequity. How a senior person feels about a student can play a powerful role in the student's future, yet the student may have no knowledge about these behind-the-scenes discussions. One woman interviewed, for example, who is now a highly respected mathematician, was suggested as a candidate for a position at a prestigious Ivy League institution shortly after she finished graduate school. Recommendations were solicited from faculty members at her graduate institution. Most were extremely positive; however, one senior person wrote a letter saying that she would be at best a good high school teacher, and that they must be out of their minds to think of hiring her. The candidate learned what had happened only because one of her colleagues confidentially told her about it. Fortunately she was able to make sure that in the future the professor in question would not be asked to send a letter evaluating her work. Had she not found out what had happened, her career could have been sabotaged. To what extent did her being female play a role? It is always hard to know; however, it is unlikely that a male who had just completed a respected doctorate at a prestigious institution would be described as fit to be only a high school teacher. How one is perceived plays a critical role in one's future, yet women are often perceived differently than their male peers.[17]

PROFESSIONAL DEVELOPMENT

The significance of connections to the mathematical community continues even after one has a job. An advocate—someone who can "vouch for" younger and less established mathematicians—can smooth the transition to the larger mathematical community. This is particularly important for women because they are more likely to be initially viewed with suspicion.[18] Such advocates help bring women's names to the fore not only in the context of job opportunities, but also as invited speakers at professional meetings, for important committees, as journal editors, and for awards.

In addition, the mathematical community guides one not only through the highs, but also through the inevitable lows of mathematical research. It helps by supplying both intellectual and emotional support. Karen Uhlenbeck:

> I tell my students that the most important thing, if you want to keep doing mathematics, is that you establish mathematical contacts. Even if you don't need to work with them, you're going to get depressed sooner or later and you're going to need some sort of input. . . . Whether people stay as research mathematicians or not, I think the big item is that they have some contact in the mathematics community of a personal nature. That sounds weird because mathematicians are crazy. They work by themselves, and you think of them as sitting in their room working by themselves, but every mathematician hits bad points. And how do you get over it? Somebody has got to come along and say, "Cut it out, kid." Or somebody has to come in with a new idea and hit you on the head with it. I see young people who always think they want to go to a place where there's a lot of action and a lot of ideas going on. I think the only benefit they really get from that is that they make strong relationships. Some mathematicians are social. Some mathematicians work together, but a lot don't. What happens to the people who go out and work in isolation? I think nothing, except that you're bound to hit a bad point and then how are you going to get over it? . . . There are people who sail through and nothing ever goes wrong. Normal people aren't like that.

In this sense, the mathematical community serves as a kind of protective net for those walking the tightrope of research. Great achievements require great risk; the net limits injury when one falls off, and gives one the opportunity to get up and try again. This is as true for men as it is for women.

It would be tempting at this point to argue that while being a mathematician may be social, the actual doing of mathematics occurs in isolation. That is, connections can be helpful in all the external ways mentioned above, but when it comes to actually working on mathematical problems, one is essentially alone. But in reality this division is not nearly so clearcut. Doing mathematics involves discussing new ideas, new problems, and one's current work with other people. Often there is not a simple dividing line between what is done alone and what is done with others. For example, Fan Chung describes the role of the renowned mathematician Paul Erdos in her mathematical activity:

> Erdos has great intuition. When I'm trying to prove a theorem he will immediately say "No," that I'm not going in the right direction, and after a while he'll push you in the right direction. Intuition is very important; otherwise you will try so very hard to prove something which is not true. It saves you a lot of energy, and he's great at pushing and pushing. When you finally think you're done, he's asking you another bunch of problems.

Marcia Groszek echoes this theme when she describes how important ideas for her thesis (and later research) grew out of ongoing mathematical discussions with a few close colleagues and mentors:

> I got academic support from a combination of people when I was writing my thesis. Gerald [Sacks—a professor at MIT] got me going in the right general direction. Ted and Peter [fellow graduate students at Harvard] gave me the ideas and questions that became my major theorem. Ted and Aki [Kanimori, an assistant professor at Harvard] gave me literature to read. It was Ted that I talked to on a day-to-day basis. I would wander over into the next alcove and tell him what was not working and what was working.

In the end she says, "I think working inside the mathematical community is really key. It's the place where I get questions, appreciation, and can share what I've found. I've had a lot of help and support from different people."

On a practical level, collaboration can be extremely productive. For not only are the relationships rich in feeding mathematical excitement, they also serve as a kind of living library. As Fan Chung, who has co-authored more than 170 papers, says, "It's a wonderful relationship. It's a little more than just friends. . . . My co-authors are my best teachers. You really learn how to actually use them. You learn a lot of proven theorems, known results, as well as how to actually use those results. You really see the action when you collaborate with other people."

This kind of intellectual interaction with colleagues plays a significant role in the lives of many male mathematicians as well. Dr. Leonard Adleman, a distinguished mathematician at the University of Southern California who helped develop a completely novel way of sending secret "unbreakable" codes, said that his best-known work grew out of an intense collaboration with two colleagues. Indeed, the problem that made him famous was one that did not strike him as particularly interesting. However, as he says of these two mathematical friends, "They would talk endlessly about it and because we were all together so much, we would discuss it."[19] That kind of social interaction was critical to his professional accomplishments.

Even Andrew Wiles, who is one of the most extreme cases of relative isolation, had one colleague (who was sworn to secrecy) with whom he spoke regularly about his ideas. And many of the important tools Wiles used in his proof were developed by other contemporary mathematicians. Had he not known about their work, he would not have made the profound breakthroughs that he made. Despite his desperate desire to solve Fermat's Last Theorem completely on his own, in the last stages of his work he relented and brought in a collaborator: "I was very tired. I'd been working very hard, and I needed someone to check every statement I made. I needed someone to talk to all the time." He had the option of tapping the mathematical community whenever he needed or wanted to. Thus, while the ideal of isolation has a powerful hold, it rarely is an accurate reflection of how mathematicians work. Nor is it clear that it is beneficial to productivity.

Given how important these connections to community are, it is not surprising that almost all of the women interviewed had formed a network of colleagues who contributed to their success.[20] The reality, then, is a far cry from the image of a loner cowboy.

These findings are corroborated by other studies in science more generally. Mary Frank Fox, for example, documents that collegial connections are one of the best clues to productivity. Her findings also indicate that men, in general, have more access to collaboration than women.[21]

The Impact of Imagery on Women

One problem, then, with the image of the loner mathematician is that it can hurt those who take it too literally. It masks how important one's ties to the mathematical community are, and can be detrimental therefore to those who believe it—those who do not develop and seek out such connection.

However, even when women do recognize the need for such connection, it is often more difficult for them to create it. The very fact that they are women casts them as "others," a hurdle they must constantly jump over with new people in the community. It is harder for women to develop an easygoing professional relationship that is also comfortable personally.[22] As Marcia Groszek said, "Participation in conferences is half professional and half social. It's not as easy to fit in and socialize if you're only one of three women. I often feel insecure about whether I should invite myself along on this expedition. Is it okay to go with this group? Do they really want me?"

Furthermore, establishing and maintaining these connections with the mathematical community takes a great deal of time, energy, and money. It involves traveling to meetings and conferences, visiting colleagues to exchange ideas and collaborate. Women often have less access to both this time and money because they are more likely to have heavier teaching loads and to be at less prestigious institutions with fewer resources.[23] Moreover, those who have family commitments have less freedom to travel, at least for periods of time.

In addition, women are less likely to be given the kind of guidance that emphasizes the need to develop strong ties with the community. Indeed, often women come to recognize the importance of community only after they find themselves outside it.[24] And those who *have* established strong ties typically describe them as accidental, something that they fell into.

Men may not recognize the role of community either, but they are more likely to reap the benefits of it automatically. Indeed, it is so much a part of the natural course of events for men to be tied into the mathematical community that it is easy for them to take it for granted, not even recognizing the way in which community is invisibly operating behind the scenes: in helping them to get jobs, in brainstorming sessions with colleagues about ideas, in the exchange of news about recent theorems in bars or bathrooms, in the suggestion of each other's names as speakers or journal editors. As one mathematician pointed out, "How do people get picked to be editors of journals? Well, there's an awful lot of contacts and friendships, no question about it. Women miss out in that. They don't have those connections."

When women have difficulty penetrating the social fabric of the mathematics community, it is easy for them to experience this as a vague sense of not fitting in, which can be internalized as self-doubt and can have a negative impact on mathematical productivity. Often women assume that since they are supposed to be autonomous, independent, and self-sufficient, any problems they face must be their own. That feeds a kind of isolation that creates a vicious cycle of declining productivity.

WHY DOES THIS IMAGE PERSIST?

If the image of a mathematician as a loner has negative consequences for women in the field, and also serves as a deterrent for others who might otherwise be interested in pursuing mathematics, why is the mathematics community not more actively engaged in transforming this image? What purpose does the image serve?

Perhaps the most compelling argument for isolation (or more specifically the focus on individualism) is that it preserves the freedom and independence that are seen as critical to creative activity. By working alone, mathematicians are free to go wherever their creative instinct and intuition guides them. As the mathematician Corrado Serge writes, "We are the only free creatures in a world of slaves. In that spirit of independence we set off into the world of reason, believing nothing because we are told it on good authority, doubting, questioning in the name of truth, leaving no stone unturned, not even the integers and the fractions."[25]

Though this argument is compelling, it is not clear that there is a direct correlation between autonomy and creativity. Certainly there are notable historical examples of tremendous achievements that were made possible by the collective effort of creative minds; perhaps the most dramatic is the Manhattan Project. This is certainly not meant as an argument to eliminate autonomy, but simply to emphasize that there are many paths to creative and productive mathematical research. However, mathematical training is still largely shaped by the belief in autonomy and individualism.

There are additional reasons why the image of the mathematician as a loner continues. The image of a devoted mathematician living a kind of exiled isolation feeds the ideology of mathematics as a heroic life. Sacrifice comes in the form of renouncing social ties for pure devotion to the pursuit of mathematical ideas. There is a kind of romanticism associated with such a monastic life.

Moreover, the image of self-sacrifice functions as a way to separate what some might view as "serious" mathematicians from the "lesser players." If social concerns are unduly important, then one does not fit the template of a mathematician. Indeed, mathematics has been described as a jealous lover, requiring complete focus and loyalty.[26] Those who are not willing to make such a commitment may be dismissed as lacking the mettle to compete in this realm.

The image of a mathematician as a loner is perpetuated by yet another ingredient: the large schism between the mathematical and non-mathematical world. Most mathematical dialogue takes place within the mathematical community. The language and ideas are so technical and highly developed that it is completely non-trivial to translate mathematical discussions into lay terms. From the outside, therefore, mathematicians seem like loners because they are isolated from those outside mathematics. Indeed, this distance between the mathematics and non-mathematics world is to a certain degree embedded in the values of the mathematics community. While many people make a concerted effort to create bridges between the two worlds, there is little institutional support for it. Expository writing, which can be an excellent vehicle for this "cross-cultural" communication, is still typically denigrated as "not real mathematics," or a lower form of professional activity.

Finally, as chapter 6 will explore, the vision of mathematics as an activity carried out in complete isolation feeds the image of mathematics as immune from social forces. It perpetuates a belief that mathematics is apolitical and value-free; hence the social arteries remain invisible.

If the vision of mathematics as a life of isolation is alienating to many people, particularly women, one might ask if the degree of autonomy that is currently expected is indeed essential. Though we have seen that, in practice, many of the most successful mathematicians are highly social, the ideology of autonomy remains the dominant imagery that drives mathematical training and shapes the lives of many other mathematicians. While there is nothing wrong with doing mathematics alone, perhaps there is a problem if this is perceived as the only mode available to mathematical development. A variety of signals convey the message that doing mathematics alone is the norm: some prestigious institutions do not count co-authored papers, research grants are typically given to individuals rather than two or more collaborators, and journals often look more favorably on singly authored papers. Autonomy may be effective; however, what some might experience as compulsory autonomy robs the mathematics community of exploring other ways of doing mathematics, and perhaps robs them of potential mathematicians. Is it possible that creating more models of doing mathematics in a cooperative mode rather than an individualistic mode could increase the quality and quantity of mathematical research?[27]

This is one area where segments of industry may well be ahead of most of academia. One of Fan Chung's primary duties as division manager of mathematics, information science, and operations research at Bellcore was to "maintain and cultivate the research environment—bringing together good problems with problem solvers and theory builders from a variety of areas through activities such as seminar series, speakers and visitors, paper distribution series, and involvement in professional societies." She would evaluate her success in terms of how much collaborative activity she stimulated. This example leaves us with a question: What would it mean for department chairs, for example, to see themselves more in the role of cultivating a cooperative research environment?

Certainly collaboration is not for all people, nor all contexts, but given how significant it has been to many of these women, as well as to other mathematicians, it is worth contemplating how to cultivate it in early mathematics training—something that is currently rare. Recognizing collaboration as one of the tools of the trade could influence how mathematics is taught, attitudes about joint work and publication, as well as how meetings are run. Seeing people as potential collaborators rather than competitors radically changes how one relates to colleagues. Another person's strengths—particularly those that differ from one's own—become assets rather than threatening, complementary rather than competing. In addition, collaboration can be tremendously fun and stimulating. As Joan Birman says, "I get much more pleasure in a discovery that is done with somebody else. That interchange between two minds at a very deep level is just wonderful. . . . That kind of contact with people is very meaningful to me."

There is a great deal of variety in how mathematicians practice mathematics. There are those who, like Andrew Wiles, prefer working in relative isolation for extended periods of time, and there are people such as Fan Chung, who has collaborated with more than fifty colleagues, and whose primary focus is on creating and stimulating community and collaboration. Hence, mathematicians can in some sense be placed along a spectrum—from those who are highly individualistic, to those who are highly social. Moreover, different aspects of any given person's work might fall on different parts of the spectrum. Some people hunger primarily for social support and encouragement, but like to actually work out mathematical ideas alone. Others find thinking ideas through with another person to be more stimulating than working alone. But the image—certainly as held by those outside the mathematics community, but to a large extent the image among the mathematics community as well—captures only one side of that spectrum: mathematicians are portrayed as loners.[28]

What is clear, however, is that this image is certainly not the whole story, and is in some ways misleading. Access to the mathematical community does make a difference and can be an important ingredient in productivity, professional recognition, and success. In particular, the community serves many roles simultaneously: it can be a source of support and encouragement as well as a source of mathematical ideas and interaction.

For a variety of reasons, women often have more difficulty gaining access to central veins of activity in the mathematics community. Sometimes they do not recognize the need for this kind of connection, but even when they do, they often have more difficulty than their male colleagues integrating into the social fabric of mathematics. This difficulty is not because the women who are drawn to mathematics are particularly asocial; indeed, they eagerly embrace the opportunity to collaborate. But as women they find that it is harder to be recognized, accepted, and encouraged as mathematicians. The image of a mathematician as a loner exacerbates this problem by making it difficult to address how community functions in the development of a mathematician's life, and how it influences women.

Finally, the image of a mathematician as a loner serves to keep many people out of mathematics. For this reason alone it should be carefully examined. Is the degree of individualism that is currently emphasized indeed necessary? Is it possible to cultivate a different image of a mathematician's life?

Karen Uhlenbeck and her partner Bob

Karen Uhlenbeck

Karen Uhlenbeck

(1942–)

In asking mathematicians around the country who they thought should be interviewed for this book, almost everyone named Karen Uhlenbeck. Uhlenbeck is considered one of the top mathematicians in the country; she has received many distinguished awards, including the MacArthur "genius" award, and a nomination to the National Academy of Sciences. Her pioneering work in mathematical analysis has earned her tremendous respect as a talented and creative mathematician.[29]

One of the ingredients of Uhlenbeck's success is that she is extremely independent. She is proud of the way she was basically self-taught, pursued her own interests, and for a great deal of her professional life "didn't need anyone." Over time, however, Uhlenbeck gradually became aware that such determined individualism can lead to isolation, and can have a negative impact on professional growth. As her career developed, community became essential to success.

At the same time, for many women, doing math is not "an automatic thing." It is more difficult for women to envision themselves in such a role, and this lack of vision can have an effect on many levels. It makes it less likely for women to pursue mathematics, and even those who begin such a path are more easily diverted if they do not have a clear sense of where it will lead or how they would fit in. Consequently, women are more likely to choose alternative trajectories that present themselves. Moreover, even those women who do stay in math have trouble seeing themselves as mathematicians. As Karen Uhlenbeck says, "Even when I had had my Ph.D. for five years, I was still struggling with whether I should become a mathematician. I never saw myself very clearly." This difficulty in imagining oneself as a mathematician arises in part from the strong social stereotypes about women, as well as from the lack of role models to present alternatives to these traditional expectations.

Most striking is the fact that even some of the most talented women in mathematics—those who are clearly gifted and who love mathematics—can still feel like outsiders in the mathematics community. As Uhlenbeck said, "I'm not able to transform myself completely into the model of a successful mathematician because at some point it seemed like it was so hopeless that I just resigned myself to being on the outside looking in."

What are the factors that contribute to this feeling of being "on the outside looking in"? And how does this sense of marginalization complicate women's relationship to public awards and recognition?

Early Childhood and Early Independence

From an early age, Uhlenbeck had a healthy disregard for the social expectations of women. The roles for women in the 1950s, the decade in which she came of age, were quite limited. In the previous decade, women had been exposed to a wide range of opportunities because so many men were away at war. Their skills were needed in factories and management, and they discovered that they could be welders, truck drivers, or professional baseball players—almost everything that needed to be done was done by women. Later, in the sixties and seventies, the women's movement and sweeping social change opened many doors for women, allowing them to envision and pursue new paths. It became increasingly common for women to become doctors, lawyers, scientists, or business executives. But the dominant, and unquestioned, roles for white middle-class women in the fifties were clear: their focus was on getting married, having and raising children, and caring for the home. Women were not supposed to be active in sports, except as cheerleaders, and being intellectually oriented made one an outcast. As Uhlenbeck said, "I did not feel like I was supposed to do anything interesting except date boys. That was what girls did."

Karen, however, preferred to play football and climb trees. "I was very much a tomboy. The boy down the street and I played football and baseball for the better part of my life, right through high school. It was not a very respectable thing to do." So entrenched were these roles and expectations that even her mother, who was far from the paradigm of convention, had difficulty accepting her daughter's unwillingness to conform to gender roles.

But despite her mother's discomfort with Karen's unconventional ways, Karen's whole family in many ways paved the way for her being different. In the rural community in northern New Jersey where she grew up, her parents were noticeably different—her father, an engineer, and her mother, an artist, were Democrats and went to a Unitarian church, while most people were conservative, Republican, and either Catholic or Protestant. Her parents were also ardent conservationists. For their honeymoon they went hiking in the mountains of the West, and later, even with four children (of which Karen was the oldest), the family would go camping every summer for two weeks in the Adirondacks. This background in the outdoors was an essential part of Karen's development.

Her family provided many models of strong and independent women. Her grandmother, an imposing six feet tall, raised twelve children almost single-handedly and lived to the age of 103. Karen's mother, also extremely active and energetic, "a kind of superwoman," had a fiercely independent streak as an artist, an intellectual, a maverick in her own right. She believed in doing everything herself, and as Karen says, "I think I get a lot of my character from her."

Growing up in a rural environment helped Uhlenbeck avoid intense peer pressure to conform. She found alternative forms of companionship in the unstructured world of the country; by playing in the fields and helping in the garden, she grew familiar with birds, flowers, and trees. And she found alternative visions of the world by immersing herself in books. She read everything she could find, particularly biographies (there were very few on women, "mostly presidents' wives"), and books on math, science, and frontier life. These early influences began to shape her vision of what she wanted to do with her own life, a vision that would enable her to escape the uncomfortable expectations of who she was supposed to be.

> I was either going to become a forest ranger or do some sort of research in science. That's what interested me. I did not want to teach. I regarded anything to do with people as being sort of a horrible profession. . . . I felt that I didn't get along with people very well. I always had a lot of girlfriends, but I never had very many boyfriends; I didn't feel comfortable. I never felt like I was really a part of anything. I went my own way, without really wanting to, but I never did understand the trick of doing things like you were supposed to.

"Not doing things like you were supposed to" may have been a liability in adjusting to social norms, but it would ultimately become an asset in her pioneering research as a mathematician.

Becoming a Mathematician: "Not an Automatic Thing"

By the time Uhlenbeck finished high school, she had no idea that she would go on in mathematics. Her early schooling had not been particularly stimulating; indeed, she had spent most of her time reading novels under her desk. College, however, turned out to be a striking contrast.

Uhlenbeck's family assumed that Karen would go to college, but where she went was less important. She chose the University of Michigan because it was relatively inexpensive; spending money on a woman's education was not considered a high priority. The year was 1960.

Karen enrolled in the honors program at the University of Michigan, which provided her with an excellent education; her strong training in mathematics there was to prove very useful in graduate school. Interestingly, this honors program produced a large number of women mathematicians. Because it was so rigorous, many people dropped out of the mathematics portion, but for Karen it was a period of blossoming, both personally and intellectually. She had lots of friends, and felt more comfortable with women from the Midwest than with those from the East Coast because they seemed to her more open and friendly. There was a large contingent of New Yorkers as well. She also began to have boyfriends, most of whom also studied mathematics or science. During her junior year, she went abroad to study in Germany. The world unfolded beyond her wildest dreams. The discipline of the classes and the polished lectures (given in German) were stimulating to her. She traveled around Europe, learned to ski, went to the opera, and got to see parts of the world that were dramatically different from her home town in New Jersey. She especially enjoyed being totally on her own.

It was during college that Uhlenbeck first discovered how much she enjoyed mathematics. Her first math class in the honors program at the University of Michigan was an extremely challenging analysis course. Karen flourished and decided to switch her major from physics to mathematics. "I got to college and discovered that I could do mathematics, and I never even saw myself as doing it, but I recognized that I was partaking in something that I enjoyed. . . . I just thought the idea of dividing things up into infinite amounts seemed really far out." She particularly liked the excitement that came from connecting with mathematics directly instead of "just doing what the book says." And she "found it really neat that you could think through these arguments and get them right just like the book." Perhaps most important, she discovered in the discipline of mathematics an incredible sense of freedom, a creative freedom not unlike that expressed by poets who work with haiku or other highly structured forms of poetry. As Uhlenbeck says, "If you obey the rules, you could do almost anything you wanted."

However, despite the fact that Uhlenbeck loved mathematics, she did not assume that she would pursue it in graduate school or go on to be a mathematician. As Uhlenbeck points out, for women, pursuing mathematics is not "an automatic thing." And she was no exception. She simply could not *imagine* herself in such a role. Ultimately, it was the fact that her boyfriend and other undergraduate friends were going to graduate school that drew her into it—an early indication of how significant ties to a community can be. But she still had no vision of where she was going or what steps were involved.

> I think if someone from industry had come and interviewed me and wanted to hire me, I might have been inclined, because I don't think my parents were particularly enthusiastic about my going to graduate school. No other option came up and said we want you, and I was being pressed by people in math. All the people I knew were going to graduate school. And I met my future husband as a senior and he was going to graduate school (in bio-physics). . . . It wasn't something that I had been geared up to doing all along. It was just that my life grew into that, especially when I started seeing him seriously. He was going to graduate school, so what was I going to do?

Karen received several fellowships, including a National Science Foundation and Woodrow Wilson Fellowship, which made it easier for her to pursue mathematics. Not only did it provide the finances to go on, it also encouraged her psychologically. While she couldn't imagine herself as a mathematician, the fellowships at least implied that other people thought she had potential.

Despite the fact that she had received highly coveted fellowships and that her boyfriend was going to Harvard, Karen never thought seriously about applying to the most elite schools, such as Harvard or MIT (Princeton was still not admitting women). She chose instead to attend Courant Institute in New York City, which has produced a number of women Ph.D.'s. Although this was a strong program, its focus was on applied mathematics, which was considered less prestigious then the "pure" math focus of the Ivy League schools.

When Uhlenbeck married, she decided to transfer to a graduate program closer to her husband. Again she did not apply to Harvard or MIT, but instead chose Brandeis. It was not so much a lack of confidence in her mathematics, ability as a sense of how problematic the atmosphere would be for her as a woman.

> It was self-preservation, not lack of confidence. I was pretty sharp, without being conscious of it, of how difficult things were for professional women. (I knew all about being socially awkward!) At the time, I may have thought that if I were brilliant enough, I would succeed at Harvard. Now I do not believe that—I believe the social pressures of surviving in an environment that would question every move would have done any woman in, unless she were particularly interested in the combat. I knew I was not interested in the battle of proving social things, so I (wisely in retrospect) avoided it.

At this point Karen had no real role models: no other women with whom she could identify, and who could help her envision herself as a mathematician. Transferring to Brandeis did help, however, by at least providing her with young role models. "For someone like me, Brandeis, was a super place to be a graduate student because the faculty were all extremely young. My thesis advisor was barely over thirty years old. . . . It was exciting to hob-nob with the junior faculty in my last year." Brandeis was just beginning a new and rigorous graduate program modeled after Princeton's. Karen was in the first year of the new program. Many graduate students with weaker backgrounds had trouble surviving, but Uhlenbeck's training at the University of Michigan and her one year at the Courant Institute served her well. For such a small school, Brandeis produced many strong graduates, and women in particular.

During graduate school, Uhlenbeck still had no clear vision of where she was going with mathematics, or even what subject area she would pursue. "I think that my career was more marked by a wandering. . . . I would decide that I really liked something and I'd be kind of bored with it by the end of the course, and then I would decide that I really liked something else and I would get bored with it. It sort of wandered around." In the end, however, this wandering was good training, for her research crosses the boundaries of many mathematical fields.

In choosing an advisor, Karen based her decision on what mattered most to her at the time: who was doing what seemed like exciting mathematics. Dick Palais was working on material that seemed new and different. Indeed, it was his field and his clarity that drew her to him; personal compatibility was less important.

> I thought he was an extraordinarily clear lecturer. I still remember I went in and said, "Tell me about the heat equation," and I got an hour lecture. That was all I needed to know about the heat equation for twenty years. He really is very clear. For me, I'm much muddier, and I appreciated that kind of a teacher. So he was an extremely good choice for me.
>
> For a long time we were very uncomfortable with each other. He frankly admits that he didn't want to take me as a student when I appeared. I think his initial reaction was that I would just have children and give up. But in fact after me he had a lot of women students. I don't think [his response] was a personal one, it was just an automatic response of the time. He wasn't negative. But I didn't choose him because I got along really well with him. I liked the kind of mathematics he did, and he was really a clear lecturer.

Thus, Uhlenbeck's graduate years were characterized by an almost paradoxical combination of traits—her lack of vision about where she was going, indeed the almost accidental way she stumbled into mathematics, alongside a strong independent spirit. She pursued what she wanted, with whomever she wanted, and with little guidance from external authorities. Even in her thesis work, she chose a subject that her advisor knew little about, so that in the end she was able to teach him some of the material, rather than the other way around. But both this tentative commitment to mathematics and her highly individualistic nature contributed to the problems she began to face as a young professional.

Beginning Life as a Professional:
On Her Own and Alone

Up until this point, being a woman did not conflict with her role as a student, particularly at Brandeis, where there was a significant presence of women. If there was any discrimination, she was ("perhaps stubbornly") oblivious to it. The clearly structured roles of student and teacher made it easy for Karen to plug into student life as both an undergraduate and a graduate student. Though she was unusual, she did not feel like an "outsider," or what she would later describe as being "other."

But in going from being a student to being a professional, Uhlenbeck found her life changing in important ways. Being a woman was to play a significant role in how she saw herself and how others saw her. She was beginning to hit territory where very few women had gone before. She became more of an anomaly, and social expectations of what it meant to be a woman would begin to conflict with her role as a professional.

It is this stage of her career that also begins to demonstrate the significance of community in shaping the life of a mathematician: what happens when there is a lack of professional community, the different forms of community that are important in one's professional development, and the unexpected ways that community can sustain one through difficult personal and professional times.

When Uhlenbeck finished graduate school in 1968, she followed her husband (a bio-physicist), Olke Uhlenbeck, for two years, taking a one-year appointment at MIT and then a two-year appointment at the University of California at Berkeley. When she was offered a permanent appointment at the University of Illinois at Champaign-Urbana, her husband agreed to move there, rather than to Princeton or Palo Alto, where she would not have a good job. Though it was unusual for a man to take his wife's career so seriously, it was beginning to be seen as prestigious to have a professional/intellectual wife, and he was proud of her in that way. But even as they followed this non-traditional path, traditional norms and expectations still had a powerful effect on Karen's life. Though she was a full-time faculty member, she was perceived, and perhaps perceived herself, primarily as a faculty wife. She did not yet have a clear vision of herself as a professional.

> I felt like I was in a cage. I did not like being a faculty wife. I remember that feeling very well. We mostly socialized with people in my husband's department. I remember eating dinner in the faculty club one time, and I went in the ladies' room and cried. I really didn't feel at home.
>
> I had an assistant professorship, but I was still trying to be a good wife. Maybe I thought of myself as that way. It's so hard to know what is going on, but it didn't work at all, and I think it was partly professional. I think it was much more professional than I realized at the time. I was very good friends with my officemate and one or two others in the faculty, but I didn't feel like it was a place where I could live. At that point, I really didn't know what I was doing, mathematically or personally. I didn't like teaching that much; I never saw it as a career. And I was trying to work by myself, really in isolation.
>
> It never occurred to me that this was not the place for me. I always did what people expected of me and kept some part of myself for what I really wanted to do. I've never been one to fight external battles. It's a waste of time. [At Champaign-Urbana] I really didn't have anybody. I did get discouraged. Before that I had always thought that I could overcome all the obstacles.

Personal and professional issues are intimately entwined for women. Even Karen had difficulty determining how much of her unhappiness was tied to her role as a wife, and how much of it came from dissatisfaction with how her mathematical life was going. Moreover, all the signals of the time were suggesting that women's lives are about being wives, mothers, and, if they are professionals, at least something like teachers. But these roles did not mesh with Karen's orientation. She was not inclined to play the role of the "good wife," she did not particularly like teaching, and because they did not have children, she was not defining herself as a mother. But she had not yet developed a strong enough identity as a mathematician to fall back on that for a sense of strength and meaning in her life, and to use it as a way to interact with the community. The cage she felt trapped in by these personal issues permeated her math life, and so she felt caged mathematically as well. At the same time her career unhappiness may have contributed to the personal unhappiness.[30]

Karen and her husband ultimately split up. In 1976 she decided to leave the University of Illinois at Champaign-Urbana and start a new job at the University of Illinois in Chicago, after a semester at Northwestern University. "That was a good move. I never had any doubt. . . . It was hard to decide whether it was professionally better or personally better. I think the city seemed like more freedom." During this period, Uhlenbeck's relationship with her professional community began to change. For one thing, it was the first time she was on her own and needed to support herself economically. Fortunately she received a Sloan Fellowship, which played an important role in the transition she was going through. "It's possible that lots of people's careers wouldn't survive if they didn't have some sort of support during a bad period in life."

> [Having the Sloan Fellowship] made a big difference to me during this period because I think I always felt that I owed something to the profession. When your personal life gets all shot, you're glad that you have something professional. That's when you suddenly realize that this isn't a game, that studying mathematics and going along and doing the next step is for real, and if you didn't have a way to support yourself you would be in a very interesting position.

Indeed, this period was the first time Uhlenbeck articulated any sense of responsibility to the profession; she also suddenly began to see herself as a professional. This link between economics and one's attitude about one's work is significant; it is one of the factors that have contributed to women's more tentative commitment to their work. To a large degree, many of the women who pursued academic work came from relatively privileged backgrounds. Those who married often did not see their income as essential to their family's security, which gave them the option of seeing themselves as either "amateurs" or "professionals." This pattern is only now changing in fundamental ways, as even married women's salaries are increasingly seen as essential to the economic stability of their families.

Though a lot of Uhlenbeck's unhappiness during the period in Champaign-Urbana came from trying to conform to roles that were not suited to her, much of it was also due to the stifling of her mathematical development. She had access to a pen, paper, and library, but mathematics is much more than an individual pursuit. Some kind of community is essential for cross-fertilization and the sustained stimulation of the mathematical imagination. And community is what Uhlenbeck began to find both at the University of Illinois and at the Institute for Advanced Study.

Breaking out of the Cage:
Giving Birth to Her Professional and Personal Self

Leaving Champaign-Urbana was a major turning point in Uhlenbeck's life. It was to signal a significant shift, a final letting go of trying to conform to an external image of a woman's life: wife, mother, teacher. As she let go of these identities, she was free to explore and embrace different parts of herself, and she began to blossom professionally.

At the University of Illinois in Chicago, Uhlenbeck was very happy with her new environment. A primary factor was the professional and personal relationships she established on her own. She no longer spent most of her time with her husband's colleagues, creating instead a community of her own. A particularly important part of that was the camaraderie, both professional and social, that she developed with other women mathematicians.

> I had a female mentor [Vera Pless] for the first time. I don't know whether she ever realized it, but she sort of saved me. She would give me paper clips and tell me what to do in trivial situations. It's really the only time I had somebody help me out. I remember that very much—being very relieved. She helped me over all the little details of a new job in a new place where you don't know anyone. I would go across the hall and bug her at least two or three times a week. She would tell me about the people on the faculty. I don't know if I would have sought her out, but she was right there, and she was there a lot of the time.

Later she says of Vera, "I felt like we were living on the same planet anyway." This kind of intimate support is extremely important, a kind of invisible support that men often take for granted, and that women often have less access to—it helps one identify the unwritten rules of a department, a university, a professional community. It is also a bond that can traverse the boundaries between personal and professional life. For women this often happens more easily with other women.

Karen also very much identified with Louise Hay, another mathematician who had gotten a divorce, and because of that she very much appreciated her Ph.D. in mathematics, and her ability to be economically independent. For Hay, as for Uhlenbeck, the seriousness of being able to support herself became real only after her divorce. With Hay, Uhlenbeck could discuss details of her life that would not arise for her male colleagues, small but significant issues such as whether to keep her married name.

Uhlenbeck stayed at the University of Illinois in Chicago for seven years. During that time she also had visiting appointments at several research institutes, including the University of California at Berkeley, the Institute for Advanced Study at Princeton, the new Mathematics Research Institute at Berkeley, and Harvard University. Both the University of Illinois and especially Princeton brought her in contact with other colleagues she greatly enjoyed working with.

The middle year when I was at the University of Illinois, I got invited and went for a year to the Institute of Advanced Study, which was a special year in differential geometry. Shing Tung Yau, Rick Schoen, Leon Simon, and J. P. Bourguignon visited. I learned a lot of mathematics that I hadn't learned, of a different kind, and got more in the mainstream of mathematics. I remember that it took me a few months before I would talk to anybody. I felt very much out of it when I came, but after I was there for a whole year—I worked with Rick Schoen during that period of time and during that summer—that was really the beginning of my success.

This group is one she continued to work with for a long time. During this period a graduate student from Harvard, Cliff Taubes, also came down to work with her, and thus began another important mathematical relationship. As she says of this period, "I don't think I became a better mathematician, but I became better able to give seminars and could say things that weren't totally incomprehensible to everybody."

Although Karen had always had some contact with other mathematicians, relatively speaking, she worked quite independently. This limited math isolation, in combination with what Karen would describe as her "messy thinking," made it difficult for her to communicate clearly with other mathematicians. She describes her interaction with a young mathematician at the University of Illinois named Jonathan Sacks, who would beat her door down until he understood what she was saying. "I was hard to understand. I still am hard to understand. I was not socialized." The year at the Institute, therefore, was very useful in teaching her how to communicate and work with other mathematicians. It was the beginning of her transition from a kind of monkish math life to working, communicating, and interacting with others in her field. In addition to teaching her how to communicate, this exposure helped steer her to "mainstream" problems in her field. In the end, this interaction was nourishing for her intellectually as well as emotionally.

It was from this group, Shing Tung Yau in particular, that she also received the kind of support she needed to gain confidence in herself as a mathematician, and to begin to see herself as such. "I could tell that Yau thought I was a good mathematician. I don't think that had happened to me before. He was obviously extremely bright. . . . He's a very remarkable and energetic person. . . . I really credit him a large amount. I'm not saying that other people haven't tried to support me. My thesis advisor has always been a large supporter. It was more real. I could tell that Yau thought I was a good mathematician. That was hard for me to accept."

But these connections did more than simply boost her confidence and bring her into the mainstream of mathematical ideas. They also became

crucial advocates in her professional life. The significance of such advocates becomes apparent when Uhlenbeck talks about one of her two older
mentors, a woman about five years older than Karen who is in roughly the
same field:

> We still have a very close relationship. She provides real support to me.
> . . . She's a typical example of a woman who is a very good research
> mathematician but who is not recognized. She is at a technical institute. The
> only good thing about the job is that she periodically gets good graduate
> students. She is underpaid and teaches a lot. I said Yau is really the person
> that I hold responsible for my success. You know, it's true—to be really
> successful you have to be protected, and there is no way to do it any other
> way. I think about this all the time.

Her experiences have helped her recognize how important it is for
young mathematicians to make contacts in the community—something
that she missed out on at the beginning of her career. As she acknowledges, given her independent streak, this was probably inevitable. However, her more recent and fruitful interaction with younger colleagues and
students is, in a sense, making up for what she missed at the early stage of
her career.

> I tell my students that the most important thing, if you want to keep doing
> mathematics, is that you establish mathematical contacts. Even if you don't
> need to work with them, you're going to get depressed sooner or later and
> you're going to need some sort of input. . . . Whether people stay as research
> mathematicians or not, I think the big item is that they have some contact
> in the mathematics community of a personal nature. That sounds weird
> because mathematicians are crazy. They work by themselves and you sort of
> think of them as sitting in their room working by themselves, but every
> mathematician hits bad points and how do you get over it? Somebody has
> got to come along and say, "Cut it out, kid." Or somebody has to come in
> with a new idea and hit you on the head with it.
>   I see young people who always think they want to go to a place where
> there's a lot of action and a lot of ideas going on. I think the only benefit they
> really get from that is that they make strong relationships. Some mathema-
> ticians are social. Some mathematicians work together, but a lot don't.
> What happens to the people who go out and work in isolation? I think
> nothing, except that you're bound to hit a bad point, and then how are you
> going to get over it? If you're on good terms with your thesis advisor, you
> call your thesis advisor up and the thesis advisor says to cut it out or gives
> you some feedback. [On the other hand] I don't want my students doing
> that. They need to find their own relationships. . . . There are people who
> sail through and nothing ever goes wrong. Normal people aren't like that.
> We have all sorts of awful things going on.

For women and minorities, this is particularly important, not because they necessarily have more problems, nor even necessarily different problems, but they often have less access to the support systems that help mathematicians through the tough periods. For those who feel more isolated, small obstacles can become enormous. Colleagues and friends can help put them in perspective.

We see, then, the subtle ways that the mathematical community can be very important in the development of one's career: it exposes one to mainstream problems and new ideas, teaches one how to communicate with other mathematicians, instills confidence, provides support during difficult periods, and facilitates recognition and professional opportunities. We also see the many levels of mathematical community. There is one's immediate work environment, and as Karen's life illustrates, through her different experiences at the University of Illinois in Champaign-Urbana, the University of Illinois in Chicago, and later the University of Chicago, these communities can fundamentally affect one's image of oneself, one's vision of the math community, and one's relationship to that community. In this way they can profoundly influence one's image of oneself as a mathematician. Karen had the opportunity to go beyond these immediate communities and meet other mathematicians who worked in fields close to her own. The Institute for Advanced Study brought together top researchers in her field and greatly expanded her interaction with the larger math community. This exposure to Yau's group was profoundly influential in the development of her career, leading to public recognition and prestigious awards.

Recognition and Alienation

Though Karen was happy in Chicago at the University of Illinois, a number of factors made her decide to leave: money was limited, she wanted to be working with graduate students, and finally she "wanted her career to go somewhere." She accepted a job at the University of Chicago, which was more prestigious; there was more money available for research, and the school had a first-rate graduate program. But in retrospect she says, "Leaving the University of Illinois and going to the University of Chicago was probably a mistake."

At the University of Illinois she had had a number of close colleagues with whom she could work, mathematicians who were not necessarily directly in her field but were close enough that they had a lot to share and teach each other. They were also young, and she found them easy to relate to. The University of Chicago, on the other hand, was not a stimulating research environment for her; she did not find colleagues to work with— "It wasn't congenial there."

The university was going through a period of transition at that time. The older generation of well-known mathematicians had all retired, and there was a younger group who were not yet as well established. The graduate program had not had much success with women students. But the mismatch between Karen and the University of Chicago went deeper than these issues. The atmosphere and traditions of the university were alien to her on many levels.

> I simply never became friendly with people. Most were educated in fancy institutions. There was this air of real elitism. I had gotten a degree from Brandeis. I'd been teaching at the University of Illinois. But they seemed to have lived their whole lives in prestigious institutions. A lot of people there hadn't taught undergraduate courses. I'd been teaching remedial calculus to business majors. A different world. They also didn't do the kind of mathematics that I did. This is not to say those things are in any way related to each other, but it was the double thing.

These various factors—the lack of colleagues to work with, the air of elitism, a long commute—all contributed to the distance and alienation she felt from the institution.

> I just didn't feel at home, I guess that's the biggest thing. I kept telling myself that I should give it a little longer, that I should do something different, but it's difficult to know how to change it. Chicago had a bad reputation among the women mathematicians in the Chicago area—it did not have a reputation as a friendly place. I just really didn't think those things were important. I now think those things are really important.

For women especially, the atmosphere of the workplace can have a very significant impact on their work, their image of themselves, their image of the mathematical community, and whether they fit into it. In Karen's case, the University of Chicago functioned to further polarize her from a kind of traditional, mainstream mathematical community. It reinforced a vision of "them" as "other," an image she could not imagine conforming to.

Karen's evolving vision of the significance of one's professional environment, particularly for women, had a strong influence on later career decisions. She has been offered prestigious positions and declined them because she felt they would not be positive environments for her professionally. And she ultimately chose instead an offer from the University of

Texas at Austin, where she now holds a Sid Richardson Foundation Regents' Chair in Mathematics.

Teaching and Research

In the last decade, as Karen has been able to develop in ways that feel more in tune with who she really is, she has come to recognize and become comfortable with her unique vision of her work both as a researcher and as a teacher of mathematics. Her thoughts about teaching have evolved over time, blending her highly individualistic orientation with a recognition of the role that the community plays in feeding and sustaining a mathematician.

TEACHING

Karen has a fairly unusual record with respect to teaching. In her early years she taught mostly low-level calculus and finite math courses. Later she taught primarily graduate courses. She has had little experience with anything in between, i.e., standard undergraduate courses or upper-level (math major) courses; therefore many of her thoughts about teaching emerge primarily from working with graduate students.

Teaching has always been difficult for Karen. Even when she was young and planned to be "a forest ranger or a scientist," she did not see herself as a teacher. The intensive social skills involved in teaching and the ability and patience needed to explain complex ideas in a simple manner were not Karen's strengths. In fact, her quick mind in many ways hindered her ability to teach. She discovered early on that the way she learned, and the way she thought about math, did not help her at all in teaching other people. "As far as teaching goes, it really is true that it takes me years to understand the difficulties students have. I just never comprehend that you have to say something twice. It took me a long time to understand that saying something once is essentially not saying it." Furthermore, Karen was in many ways self-taught, both as a student and as a mathematician. This made it harder to know how to teach other students; and there were no teachers that she was striving to emulate. Finally, the very traits that are assets in her research—including her non-linear way of thinking—can be problematic in the context of teaching.

The way I learned is totally useless for teaching. So you have to start again. I have no love of organization. The appeal that mathematics has for me is not that I can organize it; I think many successful teachers enjoy this organization. They like getting the material in a straight line, and I think that many students enjoy that kind of presentation. In fact, when I've had teachers who do that, I often find that it is a very efficient, neat way to learn material. But the kind of mathematics I do is very sloppy mathematics. I was discussing this with a colleague who said that non-linear analysis seemed so wild or untamed compared to linear analysis, and the kind of mathematics I do is really not a very organized kind of thing. So I think I have difficulties in teaching, but that kind of mathematics is going to be like that. Students who like things to be orderly and neat would be crazy to go into this kind of mathematics, where you have to learn an immense amount, not quite understand all sorts of stuff, and put a lot of things together that are completely different. I think it's probably hard to lecture on.

What emerges from Uhlenbeck's style and experience is a very personal approach to working with her students. Since each student has a unique voice, she interacts with each of them differently. Rather than trying to find a common denominator, or to teach students in the same way, she focuses on what is different about them. "First I decided that the students were so different that the idea that one would impose an outside theory of how they should learn was crazy. Then I decided that imposing an outside theory of how I should teach was also crazy."

When I started having graduate students it was an eye-opener, because I consider myself a sixties liberal, and we were all interested in teaching differently and giving people the right ideas and not being so formal. But when I started having more than one graduate student, maybe around my fourth graduate student, I realized that theories of learning are probably just nonsense because people think so differently. My first four students were all Americans. They all went through school, they all came from a similar culture, they were all interested in mathematics, they were all interested in roughly the same kind of mathematics, and they were all completely different.

Some of them think abstractly, so much so that I don't ever understand anything they say because I write differently. Some of them were so concrete; they had to start with the simplest case and work up. There's just this terrific variety between thinking abstractly, thinking by examples, thinking concretely. The brilliant student I had who was half physicist was a very loose thinker. Some of them are very tight thinkers. They all have something very different to contribute to mathematics if you prod them enough.

They all got interested. I never had a student who didn't get interested. I can't imagine trying to get a student through a Ph.D. when they weren't interested. But they're all different. When you're going to teach kids, freshmen, anybody else . . . if there's that much variation of graduate students in mathematics, how much variation is there going to be in the general population?

Karen's primary task as an advisor, therefore, is to help students find their unique styles and contribution. As she says, "In order to be a good mathematician, you've got to figure out what you can do and find out the way you think. I don't think it's so easy. You have to find your own way of doing things."

At the same time, she also appreciates that help from others at critical times can make a tremendous difference. When her students get discouraged, for example, she reminds them that everyone gets discouraged and recommends tricks that can help: having an easy problem and a hard problem to work on, trying something different for a thesis topic, taking a vacation, reading an article if you're tired of banging your head against the wall, or giving a seminar.

But the encouragement and stimulation does not go just one way. Uhlenbeck considers her students her mathematical children; working with them is a very rewarding part of her life. Many of them have also been great teachers for her, and a major source of inspiration. Her early graduate students were particularly important; they would take courses, for example in algebraic topology, and would eventually learn more than she knew, so they could teach her, making the relationship a reciprocal one. "My graduate teaching is so successful partly because I have arranged it so that I learn from them."

Indeed, when she is asked about who has been most influential on her mathematical development, she says, "I think I've been influenced by some of my students more than anything. I had a really good, bright student who was in between physics and mathematics, and when he left Chicago, I knew I was going to miss him. So I would go over to the physics seminars." She recalls that he said of their discussions, "It's really strange, you go in there and she talks and talks and talks and you never quite understand, but after a while something happens and it's worth it."

RESEARCH

Uhlenbeck's research blends geometry, topology, physics, and analysis. At the time of the interview, it involved finding connections between mathematics and the new physics. For example, she combines geometric concepts with detailed analysis of partial differential equations to describe objects as varied as soap bubbles, black holes, and certain kinds of quantum tunneling.

She is often torn now about how to use her time: doing problems, developing ideas that she's already been working on, or learning new material which could lead to new problems and a larger picture of how pieces fit together. It is her willingness to constantly explore new territory that has kept her mathematically vibrant and alive.

> There is the pleasure of doing mathematics and this real desire to learn what the physicist Ed Witten is doing. They are pulling me in opposite directions. I would like to do mathematics, but I also want to know what's going on over there, and the two are really different. I can't leave these new ideas alone. This is where the action is, and I feel that I really have to learn all this. I'm in a field that has had a lot happen in the last fifteen years. For me this is very exciting. As one of my students said, "It's like being a pioneer and walking on some area that nobody had ever walked before."

But for Uhlenbeck, a new kind of pleasure has emerged from doing mathematics, one for which age and experience are assets.

> I still get a kick out of doing mathematics. It's harder to come by now because many of my standards are higher. But now there's a new kind of pleasure of trying to fit things together, making something match something else. When I was younger I had no in-depth knowledge of mathematics. By now I know an awful lot of mathematics, and I'm really fascinated by connections.

One characteristic of Uhlenbeck's research that is distinctive is not only the subject matter she pursues, but also how she thinks.

> [How I think] is not linear. When I write a paper, it's much better to just have the basic ideas, and then I can pick them out and fill them in. If I just write this thing that goes linearly, I get confused. I've discovered that there are basically two types of mathematicians, those that really do go from point to point and get real upset if papers aren't written that way and write their papers that way and want the lectures that way, and then there are people like me who prefer the ideas to be given and the filling in [is secondary]— it's just structured differently in my mind.
> I think some people are very surprised that a mathematician would be like this. They think of mathematics as being ordered and careful and so forth. And indeed many other mathematicians are like this. Maybe more of them are very orderly as a whole. But it's not the way I think, and it's not the way I learn. In fact, there's really no way to get into communication with modern physics without just sitting through a lot of it so that it stops sounding like garbage. You can't logically work your way through this nonsense. You just sit through enough and suddenly what they're saying seems logical and starts fitting together. It's a different language.

Awards and Honors

In the end, Uhlenbeck's unique interests, curiosity, and style have all contributed to her success. And while Uhlenbeck is delighted by her mathematical successes, the external recognition that has accompanied this success has not been unmitigated pleasure—something that is hard for many people to understand. Awards such as the MacArthur Fellowship and election to the National Academy of Sciences are coveted by most everyone in the profession, but for Uhlenbeck they raise difficult issues about her identity.

> I didn't mind being a woman doing math, not supposed to be doing it, working on the fringes, succeeding in a small way, and sort of being incomprehensible and not having many students. In many ways that was much more comfortable. I was really sort of doing it for myself. Then [when I started getting awards and public recognition] I had to make a major reevaluation of who I am. Getting the MacArthur is really sort of traumatic in some ways. I just never thought of myself in any way like that.

There are several factors that contribute to the discomfort Uhlenbeck has with this recognition. There is of course the practical issue of how time consuming such awards can be: partaking in ceremonies, giving talks, socializing with other award winners and academy fellows. But the sense of burden is much deeper and more complex for Uhlenbeck than just increasing demands on her time.

Like many mathematicians, Karen is a very private person who chose to pursue mathematics in part because it was a world apart from the public arena. "You choose to do your own thing [in mathematics], and what you do is very private and personal, and three other people in the world may understand that." That privacy is what Uhlenbeck is most comfortable with, and it is what she lost as she gained recognition. Because of the awards she suddenly became a public figure, a far cry from what drew her to mathematics.[31]

Moreover, the awards forced her to publicly see herself, and recognize that other people saw her, as a *woman* mathematician, not just a mathematician. Since so many people began to address questions of women in science to her, she suddenly had to speak for *all* women. This was an uncomfortable role for Uhlenbeck. She does not see herself as a typical woman, nor does she think she makes a very good role model for women. In fact, because she found her passage a difficult one, she does not want her story to be exemplary.

Furthermore, being cast into the role raises conflicts in her strategy for dealing with gender problems. To a large degree Uhlenbeck felt that she was oblivious to gender bias, and her way of dealing with problems was often to ignore them. But when she hears the discouraging experiences of many other women, she feels a heightened sense of responsibility. "I'm not sensitive to myself at all, but I am sensitive on behalf of younger women." In her public role, particularly when she is seen as a representative of women in math, she could no longer ignore the problems. "I think you'll find lots of older women, even though they may not say they're feminists, who will get furious [about the kinds of problems women confront]. You won't find that in young women who have any hope in succeeding. It's self-preservation." At the same time, taking these problems on can be extremely frustrating, particularly when there is little one can do about them. "Once I became a member of the mathematical elite I found it a pain . . . to be a woman. There are two choices for me. I can either ignore the fact that I'm a woman or I can become a rabid animal. . . . I don't see any in-between reaction to the situation."

In her own life, though she did to some extent "ignore the fact" that she was a woman, Uhlenbeck developed mechanisms to create an environment in which she felt comfortable. She created a niche by defining herself in opposition to the very group she felt excluded by. In this way she was able to embrace what was different about herself, and in the process to liberate herself from the need for external affirmation. Woody Allen quipped that he never reads reviews of his work because if he took the good reviews seriously and was pleased by them, then he would have to accept the bad reviews and get depressed by them. Uhlenbeck, too, did not seek out or have much investment in affirmation from the larger math community. But this strategy made it confusing to maintain her identity when she suddenly found herself being defined as "one of them."

> I probably had always been looking outward in. . . . You say "those guys" and here you are, your whole career you're looking from the outside in and saying "those guys" and then you're suddenly one of them. But you're not, because you'll never be. I found it extremely hard to readjust everything. . . . I'm not able to transform myself completely into the model of a successful mathematician because at some point it seemed like it was so hopeless that I just resigned myself to being on the outside looking in. It will take a long time, if ever, before I can see myself as being really successful because I'm so conditioned to do it because I want to do it and to get along with life.

Uhlenbeck's life is a reminder that traditional measures of success can be viewed quite differently by women. This is *not* to say that women should not get these rewards, nor that they don't want them. It does, however, make us more sensitive to the fact that even "success" for women in traditionally male fields is a complex issue. There are two sides to these blessings, and while it is nice to have external recognition, the attendant burdens and responsibilities are weighty.

Uhlenbeck's life exemplifies the fact that what hinders women in mathematics is not necessarily lack of talent. Many factors contribute to the difficulty women face in feeling like equal members of the mathematics community. While Uhlenbeck has the talent and the independence to pursue mathematics on her own, her success was significantly tied to her connections within the mathematics community. But establishing these ties can be more difficult for women. In some cases, women are more likely to be seen as faculty wives rather than professionals in their own right. In others they may just be seen as different, not fitting the image of what a mathematician or a colleague is supposed to be.

For Uhlenbeck, establishing ties to the community was also made more difficult by her independent nature, and the fact that she thought quite differently than many of her colleagues. She was not socialized to "speak the same language." While this very trait of independence is what enabled Uhlenbeck and other women to pursue mathematics in the first place, in some ways it can also later become a hindrance.

Many people have suggested that the difficulties women face in mathematics are partly due to women's lack of confidence, but Uhlenbeck suggests that the problem is more one of a lack of fit. "I somehow feel that it's not so much a lack of confidence as not feeling in the right place. Women have plenty of confidence. The women that I'm talking about can do anything. That's the problem. They're survivors of life. They know they can find something else interesting to do." For this reason, Uhlenbeck argues that many very talented women choose to leave mathematics. But even those who stay can also experience that sense of not belonging, or not fitting in. This was true for Uhlenbeck, both at Champaign-Urbana and at the University of Chicago. In different ways gender played a role in making her feel like an outsider, a sense of distance that can be further exacerbated by other differences such as race, class, or ethnic background. In the end, even someone as highly self-sufficient as Karen Uhlenbeck discovered that one's professional environment can play an important role in one's sense of belonging. As she says, "There are all these trivial things which have nothing to do with your mathematical ability [which influence your mathematical life]. Mathematical ability is such a small part."

Postscript on Karen Uhlenbeck's Research
by Daniel Freed

Karen Uhlenbeck works on a variety of problems which defy pigeonholing into a unique mathematical field. The terms "global analysis" and "partial differential equations" probably come closest, but they do not convey the importance that ideas from theoretical physics play in her work and in this field generally. "Differential equations" are expressed in terms of calculus, which was invented by Newton and Leibniz in the seventeenth century. Newton was motivated by the study of motion, particularly motion of the planets. Much of Uhlenbeck's work deals with equations whose origins lie in modern versions of Newton's theory of mechanics and gravity. Though I stress the physical origins of the equations, one must realize that mathematicians study these equations in a manner much different than physicists. Often we apply these equations in new contexts, and we use them to teach us about other mathematical structures. Such is the case with the "Yang-Mills equations," whose study occupied Uhlenbeck during much of the previous decade. Her fundamental first work on these equations enabled other mathematicians, particularly Simon Donaldson, to use them to revolutionize the field of four-dimensional topology. Her later work had ramifications in the field of algebraic geometry. Without discussing the technical details, one can still appreciate the centrality of her work, which connects on the one hand to theoretical physics and on the other to more abstract fields of mathematics. Uhlenbeck has done seminal work on many other equations as well, and her current work is in yet another new direction—the theory of "integrable systems"—and promises to yield exciting results.

*Marian Pour-El, college
graduation, 1949*

Marian Pour-El, Oberwolfach, 1970

Marian Pour-El

(1928–)

On her first day of classes in the mathematics graduate program at Harvard University in 1950, Marian Pour-El arrived early to stake out the ladies' room. Most buildings at Harvard did not have women's bathrooms, including Sever Hall, where Marian's classes were held. She finally found one—through the library and across the yard in another building. Harvard was not used to women, and women were not used to Harvard.[32]

When Marian returned to her classroom, she chose a seat in the middle of the second row. Being short, she wanted to be near the front, but the first row seemed too conspicuous. As the other students filed in and sat down, they left a ring of empty seats around her; four seats behind, four to the left, four to the right, and four in front of her were vacant. Pour-El's presence as the only woman was quite noticeable. "The whole class was around me on the outside. They did not want to be near me; they were probably as nervous about it as I was."

This pool of empty seats around Pour-El is an apt metaphor for the ring of distance between her and her professional community. Though in many ways she has really been "inside" the mathematics community, gaining access to top graduate programs, obtaining prestigious positions and tenure, and being invited to give talks around the world, she has had to adjust to certain kinds of isolation in pursuing a life of mathematics.

What enabled Marian to pursue a life of mathematics at Harvard and beyond when so few other women survived in this arena? Two factors that were critical to her success were her striking degree of independence and determination, which enabled her to envision and pursue a path that her peers found unimaginable and for the most part undesirable, and her philosophy of not trying to change the system, but rather to find a way to work within it to achieve her own ends. As she said, "I think one of the basic differences between me and a lot of the women of my generation is that they were responding to the time. They wanted permission from the social situation of the time that they should go to graduate school or do this or that. I just did it! I didn't ask anyone's permission. I didn't get it."

Taking Control of Her Life

Born on April 19, 1928, in New York City, Marian could not rely on societal approval to pursue her life direction. Predating the women's movement of the 1960s and 1970s, she grew accustomed to leading an unconventional life, both professionally, as one of very few tenured women professors of mathematics, and personally, living apart from her husband for a large portion of their married life. Having struggled alone to pursue this path, Marian is amused when she hears an occasional comment like "You have gotten where you are because of the women's movement." She responds by saying, "Thanks for the compliment, but I am older than you think."

Nor could she rely on parental support in pursuing what felt like an uncharted path. She came from a very traditional background; her father, a dentist, and her mother, a housewife, expected their daughter to marry, have children, and lead a traditional life. Marian never accepted her parents' vision of her future, and indeed was hurt by the differential treatment she and her brother received. Her parents were attentive to her brother's interests and accomplishments, but indifferent to Marian's talents and desires. Her brother received microscopes, while Marian was given dolls despite the fact that she had no interest in them. As she said, she would have much preferred the microscope set, "but they never asked me what I wanted."

Being female also had strong implications for her education. In junior high school, Marian very much wanted to take Latin, but her mother insisted that she take French instead, thinking that Latin would be too difficult and of no use to Marian in her future life. And as a young girl, Marian heard her mother talk with friends about the fact that of course Marian's brother would attend college, but Marian would go only if there was enough money left over. In those days, education for a son was seen as a necessity, but for a daughter was an expendable luxury.

So from an early age, Marian decided to take control of her own life. "I just decided I wanted to do something and I would do it." By the age of ten she knew that she wanted to have a career. "I did not know exactly what career I wanted. But with the melodramatic flair characteristic of a young child, I wanted to 'understand the universe, live completely, and contribute to society.'"[33] This independence was reflected in, and fed by, her serious pursuit of ballet. As a young girl, she performed at the Metropolitan Opera House, and she vividly remembers the thrill of the floodlights and being on stage. To this day she prefers the feeling of performance in giving talks to very large audiences rather the intimacy of small ones.

Marian also decided early on that she wanted a better education than she was getting in her local junior high school, where "armed gangs roamed the halls threatening those who interfered with their pleasure," and where, "to avoid attack, girls were not permitted to leave the classroom alone."[34] So she applied to, and was accepted by, Hunter College High School, an elite, selective all-girls' high school. The other selective schools in New York City, such as Bronx Science and Stuyvesant, were closed to women at that time. Although her mother was not terribly enthusiastic about Marian's decision, Marian attended Hunter College High School anyway. She was delighted to discover that at Hunter, Latin was *required* of all students—a subject she found fascinating, and elements of which were not too dissimilar to what she later came to love about mathematics and logic.

While Hunter College High School was academically more rigorous than most other high schools open to women, there was, nonetheless, an underlying message conveyed to all the students—their primary aim in life was to find a husband. Pour-El remembers, "I had a very different point of view. I didn't feel I had to get married. It wasn't my goal. In fact, I took German classes in college because it was still considered the scientific language." At the time, science was not seen as compatible with marriage and babies, but Marian was clear about the path she wanted to pursue, and was determined to follow it.

Though Pour-El had done very well in high school, winning a New York State Regents' Scholarship and numerous other prizes, a private college seemed out of the question, since paying for their daughter's education was not a high priority for her parents. Hunter College, therefore, seemed like the only option—Hunter was where most of her high school peers went, and it was very affordable. At the time, the fee was approximately eight dollars per semester!

But Hunter College did not live up to her expectations. It was an all-women's college, founded as a "normal school" whose primary mission was to train women to be schoolteachers. Marian was frustrated by the rigid program and the limited options available to the students. "I didn't like Hunter College from the word go! I realized I wasn't getting what I wanted, but I didn't know a way out. I decided to wait for graduate school."

Pour-El ended up majoring in physics. She had enough courses to complete a mathematics major as well, but Hunter's rigid rules would not allow a double major.[35] She also would have liked to take additional science courses, particularly in biology, but her schedule did not permit it—it was filled with other requirements that she resented. Many of the students were from lower- and middle-class families, often immigrants, so the college required "social orientation" classes in which, as Marian describes, "they tried to civilize these women from 'inferior' backgrounds by teaching such things as how to eat an artichoke." Academic resources were limited, and students borrowed books rather than buying them, since it was assumed they would not need them later in life. These are the kinds of opportunities that were standard for women at the time. Education was seen as something that would refine women, make them better mothers, and equip them to become schoolteachers. It was not set up to train them to be serious professionals in the sciences.

Nonetheless, Marian was determined to go on to graduate school. Her work at Hunter College paid off, she was Phi Beta Kappa by her junior year, and she was accepted to, and received full scholarships from, all the top graduate programs she applied to, including Harvard and MIT (Princeton was the only exception; it sent back a catalogue informing her that she would not be considered because it was an all-male institution). The fact that the other fourteen graduate programs were open to her is a clear indication of how much had changed since a few decades earlier—opportunities for women were beginning to open up. Nonetheless, the social pressure for women to conform to traditional sex roles was still quite strong.

One of the ways in which women are often discouraged from pursuing non-traditional paths, particularly intellectual pursuits such as mathematics, is through pervasive messages suggesting that such a life would ostracize them from their peers. For Pour-El, it was important, therefore, to maintain ties with friends who had more traditional values than her own. "I considered it a measure of success in my college friendships that I was chosen to be a bridesmaid at their weddings in spite of our differing points of view. This confirmed my belief that one did not have to espouse the same opinions, to bend to the dictates of society, to get along and have friends. That valuable lesson in mental nonconformity has served me well throughout my life."[36] That lesson in mental nonconformity became particularly important in graduate school, when she went from a female environment at Hunter College to a male environment at Harvard.

Graduate School: Inside and Out

Marian Pour-El was torn between going to MIT or to Harvard for graduate work, and like so many women pursuing uncharted paths, she had no guidance in choosing an appropriate program. She ultimately chose Harvard because she read a description of a logic course in the course catalogue that sounded fascinating to her. Being accepted into Harvard's graduate program with a fellowship was no small feat; indeed, she was one of the first women to complete the mathematics Ph.D. program there. She began in the physics department (as an undergraduate she didn't distinguish theoretical physics from mathematics), but quickly discovered that it was really mathematics that she was drawn to, and logic in particular. She was turned off by the impreciseness of physics: approximating functions by cutting off the tail of their infinite power series representations. She discovered that she was interested in broader, foundational questions of mathematics.

When Marian began graduate school at Harvard, she was an anomaly on many counts: as a woman, as a logician, and later as an absentee graduate student.

> I looked like any other person, but they had just assumed before they saw me that I would be hunchbacked, or crippled, and would be substituting mathematics for the female attributes I lacked. When it was clear that I was just very ordinary, they couldn't understand why I'd be so interested in mathematics. They would ask, "Why are you here?" And I would say, "Because I like mathematics." For a woman, that was not considered to be sufficient reason.

As a woman, she never had the option of just blending in with the crowd. Her advisor, Professor Hassler Whitney, encouraged her to attend the department teas and colloquium. At first she was reluctant, but by the second week she got up the courage to go. Tea was a formal ritual, served in the library on lace tablecloths with delicate china. On the day she arrived, people were walking around talking as usual, but when she entered the room, a dead silence descended. The others in the room became uncomfortable and, not wanting to look at her, stared down into the bottom of their teacups. Not knowing what else to do, she stood there doing the same thing. Finally, Whitney turned the light switch off and on to signal the beginning of the talk and said in a booming voice as he looked at Marian, "Your presence is noted." As is often the case with women in mathematics, once people got used to her, it became easier. There was, however, a period when she stopped attending the teas because colleagues began to assume that she would serve the tea and collect the dishes. Gender expectations follow women everywhere they go; the mathematics community is no exception.

The awkwardness of being the only woman in the program seriously working for a Ph.D. was also apparent in her relationship with fellow graduate students. While she was able to have fairly easy interaction with some of the older students, many of whom were World War II veterans, and many of whom never finished the program, most of her classmates found it uncomfortable to interact with a woman student. Since the boundaries between personal and professional interaction were too confusing, it was easier just to ignore her.

> The older ones had a bit more poise. The younger ones, although they were older than I, didn't know how to speak to a girl without asking for a date. The situation posed problems. So I became friendly with the older ones. But even with the older students, the social interaction was not simple.
>
> Many times we decided to go to a movie or a concert. So I would go along too, and I always wanted to pay my own way because I wanted to be able to suggest a movie too. But they couldn't accept that. They would always try to be a pace ahead of me, to pay my way before I got there. There was no such thing as "dutch" then. It was considered an insult [for them not to pay] because I was a woman. I told them that I didn't want to be considered a "woman." They couldn't understand that either.

Marian's gender also caused logistical problems for the department. When it came time to assign offices, they didn't know where to put her. It seemed inappropriate to allow her to share with a man, but there were no other women in the department. They finally compromised by pairing her with a man from India who was extremely shy and quiet; there was a divider between their desks to make it even more discrete.

Not only was Marian an anomaly as a woman, she was also nontraditional in her choice of subject areas within mathematics (at least at Harvard). She was fascinated by logic and the idea that it explored basic questions about the limits of mathematics, and what is provable in mathematics. She was drawn to the idea that one could rigorously prove that there were mathematical truths that were unprovable. Since there were no logicians in the mathematics department at Harvard, she had to study these ideas on her own, piecing together information from fellow graduate students and poring over volumes of the *Journal of Symbolic Logic*. When it was clear that logic was her field of choice, she decided to go to Berkeley to pursue her studies since there were more logicians there.

The idea to go to Berkeley came from a visiting professor who saw her working on logic in her office.

> He said, "What are you doing?" He was surprised to see a woman doing math. So I told him that I was trying to learn logic. He said, "This is not the place to learn logic; there are no logicians here. Why don't you go to Berkeley?" He wrote to a colleague there, and the next year I was out at Berkeley. I didn't bother telling anybody that I was going, when I was going

to come back, or what I was doing. I disappeared. There was a rumor among the graduate students, which turned out to be true, that if you could write a thesis that was good enough, you would get a Ph.D. no matter what.

Thus, in her characteristic way of "not asking other people's permission" to do what she wanted to do, she left for Berkeley in search of a logic community.

Marian was at Berkeley for five years. During this time she met and married her husband, Akiva Pour-El, and soon after had a child. This delighted her father, who had worried that his daughter might never marry or have children.

Although Marian was in many ways an outsider at Harvard, as a woman, a logician, and in absentia in writing her thesis, she has positive feelings about the institution. From her perspective, the Harvard mathematics department was accommodating and fair where it mattered most: in granting her a Ph.D. Despite the fact that she had disappeared for several years, when she submitted her final thesis, they took it quite seriously. And since there were no in-house faculty who were qualified to judge it, they sent Marian's thesis to outside logicians who were. The faculty organized a little seminar to learn about her work before she defended her thesis. The defense went well, and her degree was granted in 1958.

Working within the System: Benefits and Costs

Her philosophy throughout this period was not to fight the system but rather to work within it. Though there were clear inequities in her life, she chose not to focus on them. Sometimes this was because she did not notice them; sometimes it didn't seem worth the energy to take them on, distracting her from the work she wanted to be doing; but most often it was because inequity was so pervasive that people just took it for granted. As she said, "You wouldn't bring your child to work any more than you'd wear a bathing suit shopping in midtown Manhattan." It was clear what was appropriate and what wasn't, and one didn't question that. So, for example, she did not think to question the inequity of a sign posted at Harvard advertising an opening for a special instructorship at Yale, which had written in at the bottom "Men only need apply." She was looking at it with one of her fellow graduate students, who said, "If you were a man, you'd have a good chance at it."

"What was my feeling about the matter? It was one of gratitude that the Harvard faculty had thought enough of me to ink in the addition. I had expended considerable effort at Harvard to overcome an inadequate mathematical background. This remark gave me a sense of achievement."[37] As she said later, "That's the way it was. And it makes me feel that maybe thirty years from now, things will be so different. Things that we take for granted even now will seem very strange in the future."

Nor did she think to question why one of the main libraries at Harvard, Lamont Library, did not allow women in. This was frustrating during the long hot summers, for it was one of the few air-conditioned buildings on campus. But it became a more serious issue when Pour-El's first teaching assignment was scheduled to meet in Lamont Library. Even as the instructor, she was not allowed in because of her sex; and rather than break that tradition, Harvard moved the class to another location.

But Marian did not believe in spending her time harping on these inequities. She felt that she was able to accomplish what she wanted without confronting these issues head on. She did not feel they were barriers to her success.

And indeed Marian was successful as very few women in mathematics before her had been. She was offered a tenure-track job at Penn State,[38] one of the centers of logic in the country, and in 1962 she received tenure with no problem. She was a visitor at the Institute of Advanced Study for two years working with Gödel, one of the most important mathematicians of the century. She began her current position at the University of Minnesota in 1964 and was promoted to full professor in 1968. In 1969 she had a sabbatical year at the University of Bristol in England. She has given invited addresses at the major mathematical meetings all over the world, including China, Japan, the former Soviet Union, and Eastern and Western Europe. Her work in recursion theory and computability has been recognized and appreciated internationally.

It might seem from the outside that Marian Pour-El had it relatively easy. She got into a prestigious graduate program with full support, and she was not prohibited from pursuing her studies somewhere else, at her own pace, giving her the time to have a child as she was completing her thesis. She had no trouble getting a job as soon as she finished her thesis. The job was a very good one, at Penn State, which had a strong logic community, and she got tenure without any problem. She had a husband who was open-minded about unconventional marital arrangements and supportive of her career.

Upon closer examination, however, one sees that many periods were far from easy, particularly in trying to write a thesis with a young baby, and even more so during her next year at Penn State, when she was a single mother working full-time.

Isolation

While being female in an almost all-male environment did not prevent Marian from achieving these notable accomplishments, it did have its cost. The price has been a certain degree of isolation. Even after years of interacting with the mathematical community, Marian still, at times, describes herself as lonely. "I think people who interact with other people have a much more joyous experience with research than I do. I'd like to make it much easier for other people. My office is on the IMA [International Mathematics Institute] floor at the University of Minnesota. I see hordes of people each year coming to talk to each other about research. It's much more joyous [that way]."

During graduate school, the social isolation she experienced went hand in hand with intellectual isolation. Because she had very little contact with students in her own class, she worked alone on homework, studying for exams, doing research. As she recalls, "It was a lonely time. . . . I was completely outside." Later in her career she collaborated with several colleagues on research, including Ian Richards with whom she worked for ten years, publishing ten papers and a book titled *Computability in Analysis and Physics.* Nonetheless she would have welcomed more opportunities for collaboration than she had. "Alone is hard. Too much alone. I would really love to work with [more] people."

However, finding mathematicians who could fully accept her as both a woman and a mathematician was sometimes difficult. She recalls meeting Steven Orey at a math meeting early in her career: "He was a very interesting man, one of the very few people—in fact, the only person I met then—who had the capacity to talk and listen to a woman irrespective of her gender. He was the only one." Only in recent years has she had the opportunity to collaborate with a female mathematician.

The periods of her life when isolation was most acute were during the early days at Berkeley, when she first had a child, and later during her first tenure-track job at Pennsylvania State University.

If there were almost no models for Pour-El of women mathematicians, models of women mathematicians *with children* were unimaginable! Once again she was exploring uncharted terrain. There were no support structures to help balance parental and professional responsibilities; it is not surprising, therefore, that having a child proved to be extremely difficult. When the baby was born, Marian and her husband were living on a very tight budget with almost no money for babysitters. Her husband was working swing shifts and was not around to help, so Marian was responsible for all the domestic work: there were no disposable diapers, and they couldn't afford a diaper service, so she had to boil the diapers, do the shopping, the laundry, the cleaning, and the cooking, and feed the baby.

> It was depressing, particularly when I had a baby and I wasn't getting out of the circle and I had nobody to talk to except baby. . . . I remember think- ing after a while that I was not getting anything done. I was very, very despondent. And my daughter was very social, like my husband. I remem- ber seeing the baby across the way who could sit for hours playing by himself. But my daughter was different, she wanted to be entertained by me.

Marian's contact with the mathematics community at Berkeley had always been limited, but after the baby was born, she was almost totally isolated from the department. The one person she did have some regular contact with was a logician in the philosophy department who had a grant that supported her for a period of time. Unfortunately he was emotionally unstable and spent a considerable amount of time in a mental institution. Once the baby came, Marian was not able to visit him there very often.

So Marian was almost completely isolated at this point both personally and professionally. She had no family to help with the child, and almost no colleagues to talk with about her work. It seems almost miraculous that under these conditions she was able to come up with a thesis. Nonethe- less, she did.

In one fortuitous week she had a flash of insight, giving her the ideas that became the core of her work. She had never even read or seen anyone else's thesis, but somehow she felt that she had material to constitute one. That flash of insight is what saved her mathematically. In the midst of changing diapers and feeding the baby, she managed to save a piece of her mind for mathematics.

Even once she had the material for her thesis, however, getting it typed was not a trivial task; she had purposely not learned how to type, knowing that as a woman she would be asked to type everyone else's work if she was able. In a reversal of roles, her husband typed the first draft, and they saved money to hire someone to type the final draft. She could have submitted her thesis to either Berkeley or Harvard, but she decided to finish her degree at Harvard.

Marian had only a vague vision of what might happen next; she thought she would either work in industry or get a job at a small teaching college. She was quite surprised, therefore, when Penn State contacted her asking if she was interested in a full-time tenure-track position. She didn't expect such an offer, and wasn't sure she wanted it, but in the end she accepted it. Her husband, Akiva, was finishing his graduate studies in biochemistry at Berkeley, and he was quite amenable to her leaving for Penn State while he stayed—a practical benefit was that it would free him of family responsibilities, allowing him to pour himself into his work at the lab day and night.

It was assumed that Marian would take the baby. And as it turned out, that first year as a single mother proved to be tremendously exhausting and difficult. There was almost no daycare, and few other forms of support were available for working mothers. When Marian finally found a daycare center, her daughter was not able to stay there long because she had recurring respiratory problems. With a great deal of effort, Marian was able to make in-home arrangements during the weekdays, but she was often up all night with her daughter whenever she was ill. At one point Marian had to go virtually without sleep for two solid weeks.

> I was teaching. I didn't get much research done. Like everything else, I just kept saying, "If only I can get through this. . . ." Then the warmer weather came, and I knew that my husband would be there in June. But it was a very long year. At the time, you have to remember all the books said that women shouldn't work, that it was bad for the children. I didn't ask for it, but I got advice and was told, "How can you leave your home and work full-time when you have a baby? She'll turn out to be a psychotic, a truant, or a criminal." And they would point to the fact that all of these children in New York City who were roaming the streets with no parental supervision turned out to be criminals. I never forgot that; you don't ever forget that.

Though there was intense pressure to be a full-time mother rather than a professional, Pour-El relished her work life. It was interesting, challenging work that brought her into contact with a larger community: both colleagues and students. This was a welcome change from the isolation she experienced in being home with her child full-time. As Marian recalls of her first year at Penn State, "On weekends there were no classes or baby-sitters, and I must say that when the weekend was over, I was sort of glad. Monday through Friday was great. I had permission to escape."

While Pour-El no longer felt isolated by being confined to her home and domestic responsibilities, she was still outside the circle of traditional social life in the community at large. No one knew what to make of her status as "single mother." Most people assumed that she was either a widow or divorced, but they were reluctant to ask. She did not feel compelled to talk about it. Others were confused, unable to label her in relation to her husband.

> It wasn't only the men who were prejudiced against women, it was the women as well. One of the things I always have to do is try to maintain a relationship with faculty wives. Many years before women started working,

one wife heard that I was going to an AMS meeting. She said, "How can you go to the meeting alone? Will your husband let you go? Doesn't he love you enough so that he will be afraid of what will happen?" And then I would hear the same old thing, women saying how terrible they are in math. That made the person more of a woman. In fact, the comments that I got from women were often worse than any I received from a man.

These issues were particularly acute in the 1950s, which she describes as "the age of family togetherness."

In a similar way, she was often isolated, or seen as strange, as a woman at mathematics meetings. Colleagues would routinely ask where her husband was. They seemed not to hear her when she kept saying that *she* was the mathematician, not her husband. At times she was frustrated by the fact that her questions were consistently ignored because she was a woman and that comments were never directed at her. These were the times she wished she had had a husband who was in mathematics. "At least that way I could have gotten my questions answered, even if the answers were not directed at me." This theme of invisibility is one that arises for many women at professional meetings, for often they are seen as women rather than as mathematicians, and in many of their colleagues' minds, the two are mutually exclusive.

The isolation that Pour-El has experienced has come from many sources. To some extent it is tied to her personality; the very same traits that enabled her to succeed—her fierce independence and self-suffi- ciency—have also perhaps kept her isolated in her later professional life. To some extent that isolation also grew out of clear practical issues such as having a child. This prevented her from having the kind of freedom and flexibility that many of her male peers had, particularly at the early stage of her career. But finally, it is also, to a large extent, a product of the fact that she is an anomaly as a woman in a traditionally male field, and hence is always somewhat on the outside. At professional meetings, in the department, or in any professional setting, being a woman influences how other mathematicians relate to her, and contributes to that isolation.

Opportunities and Uncharted Terrain

Marian hopes that future women won't have it so hard. On the other hand, she would never trade the freedom and independence she has had for a more traditional life.

At Penn State, as she has at every stage of her life, Pour-El focused on pursuing her own path without trying to convince the rest of society that it was right. She and her husband experienced periods of separation, sometimes at very long distances, other times at shorter distances, continually making arrangements to accommodate their professional lives. Almost one-fourth of their many years of marriage involved a long-distance separation. Though this was extremely unconventional for the time, and though there were certainly periods when these arrangements were quite difficult, both Marian and her husband agree that in the end it was a very positive experience for them individually, for their relationship, and for their professional development. "Living apart can be an excellent way of strengthening the marriage. When you are living far apart for a long period, when you get back together it is like being at the honeymoon stage, you want to make things work out. We didn't attempt to achieve anything. It just happened. Learning to let go. It started out being very, very temporary, and then it turned out to be for a long time."

Akiva's positive attitude about Marian's work stems in part from the fact that he grew up in Israel, where it was assumed that women would work. Being supportive of her career is what allowed him to imagine experimenting with a variety of non-traditional arrangements and long-distance family life.

Marian Pour-El wrote an article called "Spatial Separation in Family Life: A Mathematician's Choice" about the many forms "the two-body problem" can take. She talks about its advantages and disadvantages and speaks from experience in describing four different lifestyles: long-term separation, short-term separation, commuting, and living together continuously. Experimenting with these different lifestyles has had numerous benefits for Marian and Akiva's relationship: (1) It enhanced the mutual respect they had for each other; (2) since each was enthusiastically engaged in work, it increased their inner strength and independence because they couldn't rely on each other to fill in parts of themselves; (3) it led to relative equality in family roles: domestic responsibilities became low on their list of priorities and were done efficiently and easily by either of them when together; and (4) it made their lives feel exciting and rich; their time together constantly seemed new, changing, and dynamic.

Marian has been asked to speak about these options in numerous settings as many more women find themselves in similar situations. But this kind of lifestyle is not one she recommends for everyone. She tries to counsel people individually as they struggle with these decisions, because what is appropriate depends so much on the individual, the particular circumstances, his or her desires and priorities. Certainly women today are much less likely to face overt forms of resistance to such a lifestyle among their colleagues and community.

Some of Marian and Akiva's choices were determined by external factors such as institutional discrimination. Anti-nepotism rules were common. Though it was unclear whether rules banning the employment of both husband and wife at the same institution (even if they were in different disciplines) were ever written down, it was assumed that such policies were actively enforced. As Marian said, "It was so passed on that I just believed it." Thus, whether or not these rules were written down, everyone operated as if they were. In most cases this meant that the husband got the primary, stable position, and the wife would find work elsewhere—work without pay, or in an informal, usually peripheral and underpaid, position. But in Marian and Akiva's case, it was Marian who took the main position, first at Penn State, and later at the University of Minnesota; Akiva followed, often choosing industry rather than university positions, because of these anti-nepotism rules.

Marian and her husband's willingness to lead such unconventional lives opened many other doors of opportunity. Marian was offered a visiting position at the Institute for Advanced Study at Princeton in 1962. The prospect of being at the same place as Einstein and Gödel thrilled her; it was an opportunity she knew she couldn't turn down. By this time, her husband had moved to Penn State, and he stayed there while Marian and her daughter moved to Princeton. It went so well that she was offered another year. She agreed, and Akiva moved to Princeton the following year and commuted to a job in Philadelphia. While at Princeton, Marian met with Gödel regularly. They would sit and talk about mathematics once a week at a designated time. They did not work on particular problems or try to solve anything, they simply talked about what was going on in the field, what ideas seemed important or interesting. These discussions gave her the opportunity to get used to talking out loud about mathematics with another mathematician, helping her to clarify her own thoughts and ideas.

After Princeton, Marian and Akiva moved to Minnesota for five years. Eventually Akiva's work took him to Illinois, and Marian spent a year with her daughter in Bristol, England.

The Women's Movement

Marian has complex views with respect to the women's movement. On one hand she certainly supports many feminist principles: that women have the right to choose a satisfying and rewarding life, that having children should not prevent one from having a career outside the home, and that women's life options should not take second place to a husband's career considerations. On the other hand, Marian is very proud of her fierce independence, which helped her attain these things without society's approval or the support of large social movements such as the women's movement. So she vacillates between being sympathetic to women's issues (a sympathy that grows out of her acknowledgment of how hard her life has been in certain ways) and feeling that women can and should try to help themselves even independently of such social movements.

Pour-El's life also raises the complexity of intergenerational issues. Marian is representative of a kind of second wave of women in science. The first wave began to open doors that had been formally closed to women. This work was done by many women in the early 1900s who actively strategized and organized to penetrate graduate programs, giving women access to schools they had been excluded from. These women needed to work together to effect institutional change. Once these doors were opened, other women gradually began to filter through, but the numbers were still quite small. So this second wave of women were often quite isolated. Thus they could not afford to make a big deal of women's issues, since it would have meant a kind of professional suicide. There was no one to support them in this role, and it would brand them as feminists rather than serious mathematicians or scientists. Those, like Marian Pour-El, who made it through, despite the difficulties women continued to face, often embraced a highly individualistic approach. And since they were able to overcome the obstacles, they assume that other women ought to be able to do so as well. "I never fought it. It was the way it was. I went around it. That's part of the differences between what I do and what other people do. They felt that they had to fight the system and I didn't. I went around it. I never complained."

While such an approach is quite effective in certain situations, it does not necessarily dismantle other barriers that continue to exist; it can also mask the fact that a certain amount of luck is involved in not coming up against an insurmountable barrier. Some have argued that this strategy leaves the impression that only heroic and extraordinary women can pursue mathematics, and that real equality will be achieved only when ordinary women are able to succeed just as ordinary men do.

What is clear, however, is that different strategies are appropriate for different people, situations, and time periods. For Pour-El, the time and energy involved in challenging obstacles that confront women would have precluded her ability to actually do, and succeed in, mathematics. For this reason, she felt that she could accomplish more for herself, and perhaps for other women, by putting all of her energy into being a mathematician, rather than focusing on clearing the brush along that path. Other women have developed other strategies. But in the end, all of these strategies are useful; each contributes to expanding future opportunities for women.

We see repeatedly how Marian Pour-El was able to work within the system to accomplish her own ends. She accepted Hunter College as the only school that seemed open to her given her limited resources. And while she saw education at this type of women's college as inferior, she did so well academically that she was able to get scholarships and substantial financial support from the best graduate programs in the country. While Harvard was far from a supportive environment for women, she basically turned the other cheek and pursued her own path. By going to Berkeley she gained a little more access to a logic community, and despite the fact that she was not completely "one of the gang," and despite the incredible demand on her resources of having and caring for her child, she managed to produce a thesis that earned her a doctorate from Harvard. And though many schools would not even consider hiring a woman at that time, she was offered a job at Penn State, which had one of the strongest logic programs in the country, an option open to her only because she and her husband were willing to accept a long-distance relationship, though this was considered extremely unusual at the time. Finally, rather than fighting the social forces that claimed that her child would become a "criminal or psychotic" because she was working outside the home, she discreetly made arrangements for child care (though this was very hard to find) in her many professional moves: Penn State, the Institute of Advanced Study, England, the University of Minnesota.

Pour-El's philosophy of working outside society is conveyed in the following: "I've always said that people who interact in this society are in some ways governed by it. They have to get society's say-so and then fight for it. I didn't ask for or expect society to say yes." This philosophy has had its costs—it has at times meant both personal and professional isolation. For Marian, these costs are worth it. As she said, "I don't think I could have faced myself if I didn't finish some way or another." She was hungry for a life other than what she saw as the roles prescribed to wife and mother; she was hungry for an autonomous life, and an intellectual life, one that accommodated her desire to perform, as she had done as a little girl, dancing in a ballet at the Metropolitan Opera House.

What enabled Marian Pour-El to pursue this difficult and non-traditional path? More than anything else, it is what she describes as grit, a patient determinism that found failure unacceptable. But she is careful to distinguish grit from confidence. "Determination, yes. Confidence, no. I think that's one of the things that women very often do not have—confidence."

Her story shows the tremendous accomplishments a single individual can achieve even in a community resistant to her presence. At the same time, it reminds us that more work needs to be done so that women do not need to be heroic to survive.

2

What's a Nice Girl Like You Doing in a Place Like This?

Myth: Women and mathematics don't mix.

By my senior year of college, I had decided that I wanted to go on to graduate school in mathematics. I loved mathematics and I loved to teach, so it seemed natural to follow a path that would allow me to combine these interests—I thought that by becoming a college professor I could have my cake and eat it too. I was encouraged by advisors and a Danforth Fellowship and had been accepted by several graduate programs. I was not sure, however, where to go—the decision was between a large, very prestigious program and a much smaller, more personal, yet strong program. So I went to talk to the chair of the mathematics department at the large university. Though he didn't know me, he recommended against my going there. As he put it, "You're too normal; you wouldn't fit in here." I was surprised by his assessment because it had never occurred to me to describe myself as normal. In fact, I wasn't quite sure what he meant by that—that I was a woman, that I was in reasonable physical and mental health, that I had interests other than mathematics, that I enjoyed being with people, or that I had an interest in teaching? But whatever he meant, I thought he was probably right. I probably would not have "fit in." So I chose the other program.

What it never occurred to me to question was his assumption that a mathematician couldn't be "normal." His words quietly fed the sense that on some deep level I did not belong, I was not "one of them." To a large degree, what defined me as normal was simply my sex. A man, no matter how normal he appears, is much less likely to be told that he would not fit in as a mathematician. I could have "overcome" that label if there were something obviously wrong with me as a woman—for example, if I were physically or emotionally handicapped or distinctly unattractive—for in that way I would escape traditional expectations associated with being female. But short of that, I would still be seen as an outsider.

As I interviewed women mathematicians around the country, I came to see that this incident is indicative of a larger phenomenon. Not only in society at large, but even in the mathematics community itself, there is a prevalent assumption that being a woman and being a mathematician are incompatible—one can be one or the other, but not both. Thus, if women are seen too much *as women,* if they are identified with traits associated with being female such as physical attractiveness, marriage, or mother-hood, they are less likely to be taken seriously *as mathematicians.* As a result, many women mathematicians feel the need to downplay these aspects of their lives. Indeed, when these roles are juxtaposed with mathematics, the result is seen as humorous, as the accompanying cartoon illustrates.[1] The cartoon is considered humorous because there is no hiding the fact that she is a woman; in fact, she is depicted as the quintessential woman, identified with pregnancy and birth. This incon-gruity with mathematics makes us laugh.

The assumption that women and mathematics are incompatible is also reinforced by stereotypes—a woman mathematician is often depicted as an intellectual schoolmarm, someone who is assumed to be unattractive, who never had the opportunity to marry or lead a conventional life, hence who turned to mathematics as a consolation prize.

Although the reality is a far cry from this stereotype, the *belief* that women and mathematics don't mix has repercussions at all levels. Not only does it perpetuate the myth that mathematics is a male domain and discourage women from pursuing mathematics, it also affects the lives of women *in* mathematics—in how they are treated, in how they think of themselves, in their relationships with their colleagues, and in practices and policies of institutions. Dismantling this myth will entail transforming our images and expectations not only of women, but of mathematicians and mathematics as well.

Women and Mathematics: Is There an Intersection?[2]

Masked Contradiction: The Woman Mathematician

A joke that was popular among mathematicians in the past few decades provides insight into assumptions about women and mathematics: "There have been only two female mathematicians. One was not a woman; the other was not a mathematician." The two mathematicians referred to are Emmy Noether and Sofia Kovalevskaia.

Emmy Noether was a highly respected German mathematician born in the late nineteenth century. Her father was an eminent mathematician, and she worked with many of the best mathematicians in Germany, a mathematical hotbed of the world. She came to the United States in 1934 seeking asylum from the political turmoil of her country and became a mathematics professor at Bryn Mawr College. Her mathematical work in algebra was of seminal importance. "Noetherian Rings" are a mathematical structure named in her honor. She was deeply involved in the development of an axiomatic approach to mathematics. Yet in certain ways, her acceptance as a mathematician was facilitated by the fact that her colleagues did not see her as a typical woman. She was often referred to as "Der Noether," a masculine title. And in a caring and respectful memorial, the famous mathematician Hermann Weyl wrote, "The graces did not stand by her cradle." It is a strange comment for a mathematician's eulogy. How often does one see a memorial of a male mathematician that says, "Although he did good mathematics, his physical appearance left something to be desired"? Yet in her case it was considered relevant and important. It in some sense "explained" her success and acceptance in mathematics.

Sofia Kovalevskaia, on the other hand, fit traditional visions of woman-hood: the nineteenth-century Russian mathematician was physically attractive in a conventional sense, she was married (albeit under unusual circumstances), and she had a child. Yet despite the fact that she won the highly prestigious Prix Bordin Prize, which top mathematicians from around the world competed for, and that she worked with one of the best mathematicians in the world, completing not one but three doctoral dissertations, and despite the fact that she later received a professorship at Stockholm University—something almost unheard of for women—she, as the joke conveys, was still not completely accepted as a mathematician.

Women were in a double bind. If they devoted their lives to mathematics, observing the priorities of their profession, they were judged harshly as women. If they embraced more typical responsibilities or roles by having a family, they were not seen as serious mathematicians. Thus, what at first might seem like progress, namely having this separate category of "woman mathematician" which at least acknowledged some intersection between the two spheres, really masks an underlying continued contradiction. To succeed in this category, one had to be exceptional as *both* a woman and a mathematician, but the very norms of success in these two categories were in conflict. It was virtually impossible to be worthy in both realms simultaneously.

The tension between one's role as a woman and one's role as a mathematician derives in part from the fact that mathematics has traditionally been identified with the realm of mind, while women are traditionally associated with bodies, children, hearth, and home. Moreover, women are typically cast in one of three roles: wife, mother, or romantic mate. Yet none of these roles is seen as compatible with being a mathematician.

The tension between these two spheres has a long history. In the nineteenth century, there was a common belief that "as the brain develops, the ovaries shrivel," implying that women's participation in the life of the mind would impair their ability as mothers. Intellectual pursuits, particularly mathematics and science, were seen as men's sphere, and domestic responsibilities were women's. The two spheres were seen not only as separate, but also as hierarchically ordered: the life of the mind was considered far more important than the life of the home.[3] Plato did not even consider the possibility of women's leading a life of the mind. Kant continued in this tradition, defining mathematics as the realm of men, saying that women should not worry their pretty heads about geometry—that they might as well have beards.[4]

In spite of the perceived incompatibility of the two spheres, however, a few women did manage to cross this cultural boundary.[5] One might think that these women would prove that mathematics and womanhood were *not* incompatible. However, their presence did not eliminate these deep-seated assumptions. These women were usually seen as aberrations; most "ordinary" women, it was still assumed, were neither able nor inclined to compete in the intellectual sphere. Moreover, despite the accomplishments of these exceptional women in mathematics, they were always judged by traditional standards of womanhood. Failure to be spectacular in *both* camps was further evidence that the two spheres were incompatible.

Why Can't a Woman Be More Like a Man?

The tension between expectations of women and expectations of mathematicians is not simply an artifact of the past. It continues to exist today, albeit in a subtler form. The belief that one can be a woman or a mathematician, but not both, is compounded by the fact that women mathematicians are typically seen first and foremost as women and only secondarily as a mathematicians. When a woman mathematician enters a room, attends a meeting, goes to a conference, or applies for a job, the first thing that is noticed is that she is a woman. Later, especially with colleagues she works closely with, she can come to be seen as a mathematician, but that is not the default assumption. As a result, many women talk about feeling "guilty until proven innocent."[6]

Judy Roitman describes how startling it was when Ken Kunen came as a visitor to Berkeley where she was a graduate student: "He was treating me completely different than any of the other faculty members at Berkeley had treated me—like a person, not like a woman. And like a woman doesn't necessarily mean sexually. It means just that I have been removed from their category. I am in the category of 'other' and have to somehow be treated differently."

Marian Pour-El describes her experiences as a graduate student at Harvard:

> I looked like any other person, but they had just assumed before they saw me that I would be hunchbacked, or crippled, and would be substituting mathematics for the female attributes I lacked. When it was clear that I was just very ordinary, they couldn't understand why I'd be so interested in mathematics. They would ask, "Why are you here?" And I would say, "Because I like mathematics." For a woman, that was not considered to be sufficient reason.

In this way, women are often expected to *explain* their presence; it is not seen as something natural for them to do.[7]

However, in order for women to be accepted as mathematicians, it is not as simple as their simply acting more like men—for when women assume the traits that are accepted (and expected) in male mathematicians, they are met with a very different reception. A classic example is women's relationship to ambition. While ambition is considered normal in men, it can be seen as offensive in women. As the mathematician G. H. Hardy writes, "A man's first duty, a young man's at any rate, is to be ambitious. Ambition is a noble passion which may legitimately take many forms."[8] However, this is an example in which *man* does indeed mean "male," and is not being used generically; for a woman, ambition is not considered a noble trait. As Cathleen Morawetz, the director of the Courant Institute, said in a recently published interview:

> Until the women's movement of the late sixties it really was considered very bad form for a woman to be overtly ambitious, very bad form. Everybody thought that way—my colleagues, Herbert. It was fine for me to have a career, but not actually to show my ambition. Although it was positive to say that a man was ambitious, you could never say it about a woman—except negatively. And I think of course that underneath I was always very ambitious.[9]

One might think that, at least now, these issues have disappeared. But the interviewer's response to her comments is quite telling: "I never had the impression, you know, that you were out after these jobs that you have got—director of the Courant Institute and so on. Whether you cultivated it consciously or subconsciously, or whether it was just your nature, you did not pose a threat to the men here."[10] Instead of telling her that there was nothing wrong with being ambitious, he subtly praised her seeming *lack* of ambition. Furthermore, he articulated what women's ambition represents—namely, "a threat to the men." But why is it threatening for women to want interesting and challenging jobs? First, it means that they are breaking from the traditional roles ascribed to them—as wife and mother, both of which are identified with home rather than the professional sphere. Second, by entering the work force, they are encroaching on what is traditionally identified as men's sphere, and in this way pose a threat of taking away what is seen as a man's position.

This phenomenon is not unique to mathematics; a similar pattern arises in many traditionally male fields. However, mathematics sits at the nexus of two spheres, both of which are seen as disjoint from women: the world of the mind and the professional sphere. Women, on the other hand, are placed at the nexus of the counterparts of these spheres: the world of the body and the personal sphere. The classic separations between mind and body and between the personal and professional sphere reinforce the belief that women and mathematics don't mix. Furthermore, what distinguishes mathematics is its claim to being gender-blind. Indeed, because mathematics is seen as a realm of complete objectivity, it is often assumed to be immune from the kind of gender assumptions that permeate society more generally.

Because there is often little intersection between expectations of women and expectations of mathematicians, there is a very narrow band of what is seen as "appropriate behavior" for women in mathematics. Indeed, like women in many other traditionally male fields, women mathematicians walk a thin line with respect to a whole array of traits that relate to professional life: the tone of voice they use, their physical appearance, how they negotiate authority with students or colleagues, how they interact socially and professionally with peers. In each case, men have a much broader range of acceptable behaviors than those accorded women.[11] If women are too forceful in their personality or voice, they are prone to being labeled as overly aggressive or shrill. If they are physically unkempt, they are seen as messy rather than eccentric. For men, being different or eccentric can be a sign of genius, but for women it is usually accompanied by a loss of respect. One woman mathematician talked about a particularly tall student who was in the habit of patting her on the head after class: "You develop a style to put on a little bit of professional dignity to avoid that kind of condescension, and then you get called for being distant." While that kind of professional aloofness is seen as acceptable in a male professor, it is viewed as a negative trait among female professors, who, as women, are expected to be nurturing and warm.[12]

However, women not only risk falling off the tightrope of acceptable behavior by being "too much like a man," they also risk falling off if they are perceived to be "too much like a woman." Thus, for example, if a woman displays emotion, she is seen as not professional. If her voice, dress, or behavior is too far from the traditional male norm, that is reflected in her evaluations from students ("Her voice is too high") or in how she is treated and perceived by students and colleagues.[13]

Not surprisingly, the tension between being a woman and being a mathematician is often strongest in the two areas most typically identified with women's roles: that of being a romantic mate (an attractive woman) and that of being a mother. One woman interviewed described what it was like for one of her classmates in graduate school:

> J was the most attractive of the three of us [women graduate students]. She came from Smith College, a blonde, attractive woman. Very sure of herself. She kept raising her hand and answering questions because she knew them all. But she was shot down the most. I would say she was shot

down the most because she was such a threat. I couldn't believe it when people weren't taking her seriously. She was just being outspoken in class, but she was answering things correctly. I think she had the hardest time of all of us. In the end, she decided to go into another field.

In order to avoid this kind of resistance, some women talked about "dressing down" in order not to draw attention to themselves as women. One had a ritual of changing into her casual clothes at work, and then dressing up when she left. Women mathematicians rarely wear makeup or dress in clothing that might be seen as too feminine, and those who do often have a more difficult time being taken seriously as mathematicians.

Equally problematic is being seen as a mother. Many of the women who were interviewed for this book were careful to keep their personal and professional lives separate—they did not talk about their children or dwell on any aspect of their domestic life. As Marian Pour-El said, "You wouldn't bring your child to work any more than you'd wear a bathing suit shopping in midtown Manhattan." Lenore Blum did not talk about her pregnancy with anyone while she was a graduate student at MIT. Though she could hardly disguise her changing body, it did not dawn on anyone why she was suddenly wearing dresses. Indeed, her advisor was so surprised when he saw her holding a baby that he wondered whose it was and where it had come from. Fan Chung was also careful to play down her pregnancy, and was grateful that she did not have to have a formal job interview when she finished graduate school, since she was pregnant at the time. These women felt that being identified with such womanly traits as having children could undermine their being seen as professional mathematicians. As Blum succinctly put it, "I couldn't let people know I was pregnant because they wouldn't take me seriously." If men occasionally bring their children to work, they are likely to be praised for being good fathers. When women do this, their commitment to their work is likely to be questioned.

Gender expectations can have very real and practical ramifications. Several women mentioned examples of departmental attempts to take away fellowship money from female graduate students when they got pregnant. The logic behind these decisions was that since these women were now *mothers,* they no longer needed to be seen as *mathematicians.* As mothers, it was assumed that they would be supported by their husbands and did not need to support themselves. Of course that logic, in reality, is illogical, since a woman who has children needs the financial support even more (to pay for daycare) in order to continue her studies.

Because of these kinds of assumptions, women must constantly prove that they are serious mathematicians (especially if they have children), because the default assumption is that the two are incompatible. Hence in public, most women are careful to reinforce their professional side and downplay their domestic side. When Cathleen Morawetz was asked whether she didn't worry about her children when she was at work, she replied, "No, I'm much more likely to worry about a theorem when I'm with my children."[14]

Thus, in a variety of different ways, in different contexts, and to different degrees, there is still a tension between expectations of women and expectations of mathematicians. Moreover, women mathematicians are often seen first and foremost as women and only secondarily as mathematicians. But what impact does this tension have on their lives?

The Impact of the Myth

At an early stage, the myth that women and mathematics don't mix has clear ramifications: it makes it less likely that women will receive an equal education in mathematics classes, and it makes it less likely that they will be encouraged to pursue mathematics in college and beyond.[15] Sometimes the belief that women and math don't mix is hidden in the guise of what would be in the best interests of the student. Women may be told that it is not in their "nature" to pursue mathematics, or that mathematics is so completely consuming that there is no room for family or other commitments.

However, these issues do not end at the pre-college level. They can arise even once women have made a serious commitment to mathematics in college, in graduate school, and beyond. Faculty members may resist working with women graduate students, assuming that they will have babies and drop out; some have difficulty seeing women as anything other than high school teachers; and still others have difficulty seeing them as anything other than sexual beings. Furthermore, as the interviews for this book attest, women often have a more difficult time forming productive relationships with professional colleagues, and being accepted with the same kind of professional respect and recognition that their male counterparts receive. Since most colleagues, mentors, and advisors are men, the relationships are invariably complex.[16] The issue of sexuality is often an undercurrent—sometimes acknowledged, often not.

The tensions that arise from being seen primarily as a woman, rather than a mathematician, can have very real implications in women's ability to work with colleagues. As Judy Roitman relates:

I have a terrible time when I go to Eastern Europe, which I used to do quite frequently because that's a real major place for my kind of mathematics. Many of the men there are so courtly that I just can't deal with them. I can't work with them. They're just not treating me straight.

What do you mean by courtly?

It's a kind of deference that makes you "other." So that you're not really getting down to what you need to do. It's like this little kind of dance. And again, I'm not saying that they're attracted to me or that they're trying to put a make on me; it's nothing like that. It's just that I'm "other," and they cannot talk to me with the same kind of vibrancy and casualness and immediacy with which they're talking to each other.

Ironically, while the practice of seeing women mathematicians first and foremost as women makes them highly visible in one sense—as women— it also makes them invisible in another sense—as professionals. Women's invisibility has been well documented in the classroom: women are less likely to be called on, they speak not only less often but for shorter amounts of time, and they receive less of a teacher's attention and feedback.[17] These same patterns continue in professional life. At professional meetings, for example, women are often ignored, especially if they do not have an established reputation or are with a group of people who are not familiar with them or their work.[18] Marian Pour-El, for example, describes her experiences at professional meetings in which colleagues would routinely ask where her husband was, assuming that he must be the *real* mathematician. Even after she would repeatedly tell them that *she* was the mathematician, her questions and comments would be ignored.

Often these patterns are invisible to both men and women. In a classroom, they are so entrenched that often only by being carefully coded for, using precise numerical documentation, do they become visible. Thus, what feels "normal," for men, women, and teachers, is for men to speak more. Indeed, when women speak in proportion to their representation in class, it is perceived that they are dominating the class.[19] What is surprising, therefore, is the degree to which so many women at the professional level do notice these patterns of invisibility—an indication of how prevalent they are.

There are additional practical consequences to the fact that women mathematicians are typically seen first and foremost as women: they are often assigned tasks that are viewed as "women's nature." They are more likely to be identified with teaching rather than research, and are given more of the introductory courses that, as Pat Kenschaft describes, "require more hand-holding."[20] Women also tend to be assigned more service-oriented responsibilities and committee work. This is not to say that women are not interested in teaching and in this kind of committee work, but rather the *assumption* that they are perpetuates a cycle that makes it harder for women to break from those stereotypes. Furthermore, these kinds of assignments are often less prestigious and are not what is primarily rewarded in promotion and tenure.

Not only does the myth that women and mathematics are incompatible have practical ramifications in terms of how women are perceived and treated, it also affects practical choices women make in their lives. The assumption that mathematics and motherhood don't mix can lead women to postpone having children until later in their careers (which can give rise to a variety of other problems associated with age, including increased likelihood of infertility and higher incidence of Down's syndrome), or they may choose not to have children altogether. Marcia Groszek:

> Up until recently I was really not into kids—mostly because I didn't see how it could work. . . . Almost all of the colleagues I've had for role models are male. I see the way they manage to be parents; their spouse is the primary caregiver, and that seems to work for them. I didn't see how that would work for me. There are very few examples that really touched me. I didn't know any women who had succeeded being as active in research as I would like to be as well as being mothers.[21]

Beliefs about women and mathematics can affect not only how the mathematics community perceives women, but also how women mathematicians perceive themselves.

EMOTIONAL RAMIFICATIONS

Some women see the stereotypes about women and mathematics as background noise mostly irrelevant to their lives. But for others, these beliefs can have an emotional effect that can be manifest in a variety of ways: feeling like an outsider; a nebulous feeling of inadequacy, or of not quite fitting in; an increased anxiety about failure; or a kind of silencing of one's innermost feelings.

At the college level it has been documented that "when faced with doubts about their ability and their commitment, many women, not surprisingly, lose self-esteem and career confidence even though they may stay in school and earn good grades."[22] A similar phenomenon can happen with professional women. Women mathematicians can begin to internalize the views of women held by the mathematics community and respond in a variety of ways. As Judy Roitman relates:

I think because of what happened to me in graduate school (as well as other women in graduate school), it's harder for us to trust our colleagues. I think I tend to be a little pushy and strident because I had to be in graduate school to be heard. So I am generally not a very good colleague. I don't think I socialize correctly or work well with other people. Basically, I'm too sensitive to not being taken seriously, and I tend to simply fade away and give up.

In this case, Roitman internalizes the differential interpretation of one's behavior based on gender. Assertive behavior becomes equated with the negative attributes of being "pushy and strident." A man's assertive behavior is rarely described in this way. This internalization feeds self-doubt. To some extent she blames herself for being "too sensitive." And this doubt leads to a tendency to "fade away and give up."

Lenore Blum describes how many women mathematicians often join in laughing at jokes about women in mathematics because they do not think that the jokes have anything to do with them personally. As she said, often it takes many years for women to recognize the ways they have internalized these negative images of women in mathematics, further undermining their confidence in themselves.

For other women, what develops is not so much doubt about their ability as doubt about their ability to fit in. It is not that they feel inadequate mathematically, but rather that they feel like an outsider in the mathematics community. As Karen Uhlenbeck put it, "One of the most serious problems women . . . have is conceptualizing and acting upon the subtle non-articulated lack of acceptance." And this lack of acceptance can have a powerful effect on how one sees oneself as a mathematician. Even Karen Uhlenbeck, a mathematician of the highest repute who has won a MacArthur "genius" award and is a member of the National Academy of Sciences, says, "I'm not able to transform myself completely into the model of a successful mathematician because at some point it seemed like it was so hopeless that I just resigned myself to being on the outside looking in."

FEAR OF FAILURE

The internalization of negative imagery is tied to another phenomenon that is common among, though not unique to, women mathematicians, namely a strong fear of being wrong—a fear that can hinder learning. Doing mathematics involves making mistakes, which entails a willingness to be vulnerable. This fear of failure is something that many men experience as well; but for women it is exacerbated by the fact that simply because they are women, their ability is always in question. They carry around with them the residues of such doubt.

We see evidence of this fear of failure in a variety of ways, for example in women's underrepresentation in reapplying for grants after initial rejection, or in their tendency to submit articles to less prestigious journals than they might be considered for.[23]

For many women, the cost of being wrong is too high, since their ability is always being questioned. But as Fern Hunt points out, a creative spirit necessarily entails a willingness to be wrong. This has always been, and continues to be, an enormous challenge for her.

> Willingness to look bad, to be wrong, is very difficult. . . . Things would have happened a lot faster had I been more confident about that. It took a very, very long time, and I'm definitely better than I was, but I still have reticence. You can't really make good things happen unless you make a certain number of mistakes. And you might as well go through them right away.

Being a minority and a woman increases the pressure even more. As Vivienne Malone-Mayes comments: "I think we felt that we had about 11 million blacks in this country, and you thought you had all the reputation, all 11 million of them riding on your shoulders. I'm telling you, going around trying to be proof for 11 million people is a tremendous burden."

IS THERE A PLACE FOR EMOTIONS?

The image that women and mathematics are disjoint has another impact on women as well. Emotions are relegated to women's sphere, and as such are seen as having no place in the professional domain of mathematics. As Alfred North Whitehead wrote: "The leading characteristic of mathematics is that it deals with properties and ideas which are applicable to things just because they are things, and apart from any particular feelings, or emotions, or sensations, in any way connected with them. This is what is meant by calling mathematics an abstract science."[24]

But the exclusion of topics pertaining to emotion makes it difficult for women in mathematics to discuss their full experiences in this traditionally male field, including the emotional ramifications of discriminatory practices. Such topics certainly seem inappropriate for professional mathematics meetings. They are difficult to raise even in smaller contexts such as department meetings, particularly if women are in a very small minority, as they usually are. Women fear being labeled by stereotypical female attributes: "too emotional," "too sensitive."

Indeed, one of the few forums where these kinds of issues have been brought out in the open is the meetings of organizations such as the Association of Women in Mathematics. But even the AWM now struggles with this issue. On one hand they recognize the need for this kind of forum, but at the same time they worry about being sucked into a kind of "touchy-feely black hole." They worry that this focus does not help women get on with the business of being mathematicians. Some argue that the AWM should take more of a traditional focus, for example sponsoring more talks by women about *mathematics,* not about being a *woman* in mathematics. Other women don't believe in talking about these issues because they believe that one can in some sense create the problem by making a big deal of it. As a result, the topic of what women experience as women in mathematics is shunned in most mathematical contexts.

However, if women do not have a forum for discussing these issues, it is not unusual for them to internalize pervasive attitudes that women and mathematics are incompatible—they can begin to doubt their ability, feel like they don't fit in, or that they are simply unwelcome. Over time, these beliefs can be detrimental to mathematical development. Rather than seeing the larger pattern, they may decide that they are not cut out for mathematics and leave it altogether, or if they stay in mathematics, they may become more isolated and less productive in their work.

How can these patterns be changed? Clearly one step is to address the isolation that many women in mathematics feel. This involves not only encouraging more women to pursue mathematics, but developing forums for women in mathematics to find other women they can interact with professionally and socially. But another step involves changing the mental picture of who a mathematician is supposed to be. This involves increasing the visibility of those women who already are in mathematics, and developing a richer and more thorough understanding of what their lives are like.

Alternative Models

MARRIAGE AND MATHEMATICS

Contrary to the myth that women in mathematics are unmarried schoolmarms, almost all of the women interviewed for this book were married, had been married, or had long-term partners, and to a large degree these marriages were assets rather than liabilities to their professional development. What is perhaps most striking is that the majority of them were married to other mathematicians. Why do women marry other mathematicians? What are the advantages and disadvantages of this arrangement?

The most significant advantage of being married to a mathematician is that it provides a built-in community: someone to share ideas with; someone who can appreciate the fun and excitement of doing mathematics, as well as the inevitable frustrations; someone who may have ideas for a solution, or a useful reference, or simply be able to act as a sounding board. Such a partner can also be a companion at meetings, someone to travel with, or someone who can appreciate mathematical humor.

Perhaps the quintessential example of a mathematical marriage is that of Fan Chung and Ron Graham. Their mathematical bond is strong—they have worked together in industry, collaborated extensively on research, and published numerous articles together. Their life and community is mathematics. To them, mathematics is as much play as it is work—each appreciates having a partner to share it with.

Other women in mathematical marriages did not necessarily work with their husbands. For example, Lenore Blum did some joint work with her husband, Manuel, but most of their work was done independently. And Mary Ellen Rudin, though married to Walter Rudin, another distinguished mathematician, finds the idea of working with her husband totally unappealing. She likes having something that she does completely on her own. Nonetheless, they were all able to have a deep appreciation for their partners' work and enjoyed the sense of shared community.

However, being married to a mathematician has potential disadvantages as well. Two primary problems that are likely to surface are competition with a spouse and being overshadowed by a husband who is a prominent mathematician.

Some mathematicians, such as Mary Ellen Rudin, were able to deal with these issues by carving out a distinct niche for themselves that removed the possibility of direct comparison. While Rudin saw her husband as a professional, she defined herself as an amateur mathematician. He had the full-time tenured position, she was simply a lecturer who taught when she wanted to and had a few graduate students; she was not competing with regular professional faculty members for grants, awards, committee appointments, or recognition. Nonetheless, she did come to be internationally recognized as a distinguished mathematician. And because she did not collaborate with her husband, there was no question that her work was hers alone. She was sufficiently spectacular as a mathematician that she could survive and thrive on the margins.

For Fan Chung, the concern about being overshadowed by an established and distinguished mathematician had to be carefully thought through and handled. Though she thoroughly enjoys collaborating with her husband, she was very careful to do a larger percentage of her work either on her own or with other collaborators, so that there could be no question that she was a strong and prolific mathematician in her own right.

In a few cases, husbands became frustrated by their wives' successes, particularly when the wives were getting more recognition than they were. Comparisons in accomplishments were inevitable. Typically those marriages did not last.

Not all women, however, married other mathematicians. Some married men in other academic fields: Joan Birman's husband is a physicist, Judy Roitman's husband is a classics professor, Karen Uhlenbeck was initially married to a bio-physicist, Marian Pour-El is married to a biologist. The exceptions were those married to people completely outside academia: Marcia Groszek married an eclectic man who does bookkeeping and is the primary caregiver for their son. However, what was almost universal, indeed seemed to be a necessary ingredient, was that the spouse/partner was extremely supportive of his wife's having a career. The inevitable juggling, sacrifices, etc., entailed an open-mindedness to nontraditional roles.

Clearly, then, most women mathematicians are married, and usually happily so. There is no evidence of a higher rate of being single, divorced, or unhappy in a marriage. If anything, husbands seemed, in general, to be particularly pleased with and proud of their wives' accomplishments and in many cases enjoyed being able to share their life work with their partners.

CAN SHE BAKE A CHERRY PI?

While it seems plausible for women mathematicians to be married, having children is another matter. Are mathematics and *motherhood* compatible? The simple answer is yes. Many women mathematicians have had children and have successfully integrated them with their work. For some this was a smooth integration, for others it was not. What factors influence the compatibility of mathematics and motherhood? What makes it seem relatively easy for some, and extremely difficult for others? Based on these interviews and discussion with scores of other women in mathematics, the following seven factors all play a significant role:

- number of children
- the availability of child care
- the disposition of the child: temperament and health

- access to a support network: family, friends, spouse, etc.
- support and flexibility of the work environment
- health and energy of the mother
- economic resources

Those women who had the smoothest time integrating children into their work had most of these factors working to their benefit. Lenore Blum, for example, had only one child in the relatively flexible period of her career, during graduate school. She had strong support from her mother and husband, her friends thoroughly enjoyed being with her son, and he was a healthy, relatively easy baby. Similarly, Fan Chung had her first child during graduate school, her second two years later. She had excellent full-time live-in help to take care of the children, a supportive spouse who was actively involved with them, and a research position which was somewhat more flexible than a teaching schedule. She and her children were in excellent health, and she was at ease juggling many things at once.

Mary Ellen Rudin had the most children: four. But she was not in a regular tenure-track position. She worked part-time, in fact as little or as much as she wanted. Thus she could adjust her research and teaching commitments to fit her personal and family needs. She also had a tremendous caregiver who was in many ways a second mother to the children.

But for other women, integrating a child with work was much more difficult. Although Marian Pour-El had only one child, born while she was in graduate school, she had virtually no social or family support, her husband was working more than full-time, money was so tight that they could not afford child care, and they were far from family and friends. When Pour-El began her first job, she took her infant daughter with her, but her husband stayed in California to finish graduate work. So Pour-El was essentially a single parent at a time when there was a great deal of social disapproval of women's having full-time careers. Moreover, her daughter had serious medical problems which eliminated even the option of daycare. Weeks and months went by when Pour-El was barely sleeping. It took a kind of heroism to survive.

Other women whose children had serious physical or emotional problems found the tensions between work and home extreme. In all of these cases, institutional support and flexibility played an important role when it was there, and made the difficulties women face much greater when it was not.

This is an area with much room for improvement. Women should not have to be heroic in order to be mathematicians and have children.

Chapter 3 explores a variety of ways that institutional structures can help rather than hinder the integration of these roles.

In what follows, I give two more in-depth portraits of how women have mixed mathematics and motherhood. The first essay describes Mary Ellen Rudin's life. The second is a profile of Fan Chung which describes her mathematical development, her mathematical marriage, and even doing mathematics with her children. A third example can be found in Marian Pour-El's article "Spatial Separation and Family Life," which describes how she and her husband lived many different permutations of a long-distance marriage and the hidden benefits of such an arrangement.[25]

Many of the other biographical profiles in this book relate to the themes in this chapter as well. Collectively, they help dismantle the myth that women and mathematics don't mix.

Mary Ellen Rudin, 1950

Mary Ellen Rudin, 1995

Mary Ellen Rudin

(1924–)

Mary Ellen Rudin's life presents a fascinating example of how radically expectations can change in the course of one generation. Her story illustrates some of the dramatic shifts that have taken place in women's vision of themselves, their work, and their families.[26]

Mary Ellen is a kind, generous person and a very talented mathematician. She married a mathematician, had four children, and was actively involved in mathematical research. Though she is now a full professor at the University of Wisconsin, for most of her working life she did not have a standard professorship at a research institution. She describes herself as an "amateur mathematician"—a label that is perhaps misleading, for it does not convey the fact that she has published more than eighty articles in set-theoretic topology, and is internationally recognized as a first-class mathematician. In fact, her reputation in the mathematics community became so strong that the University of Wisconsin was sufficiently embarrassed by her status as a lecturer that they suddenly promoted her from lecturer to full professor in 1971. She was forty-seven years old.

At the University of Wisconsin, Mary Ellen became a kind of mathematical mother, helping to establish a strong and flourishing logic and set theory community. People from all over the world would travel there in order to be part of the stimulating, exciting, and supportive environment. She continues to be productive in publishing articles in new and difficult areas of set-theoretic topology. However, for Mary Ellen, being a mother and wife is just as important as being a mathematician. As she says, she came from the "housewives' generation." There was no question in her mind that she would have children, and that she would follow her husband wherever his career would take them.

Indeed, Mary Ellen feels that her situation was in many ways ideal—she had all the advantages of a mathematical life: a professional community, the option to teach when she wanted to, the stimulation of graduate students, the resources of a first-rate library, and an office. Yet she was not weighed down with administrative responsibilities and time-consuming meetings. The informality of her position allowed her to shape her work around her family. She could adjust to the ever-changing needs of her children and devote whatever free time she had to research. "I didn't have to prove to anybody that I was a mathematician, and I didn't have to do all the grungy things that you have to do in order to have a career as a mathematician. The pressure was entirely from within. I did lots of mathematics, but I did it because I wanted to do it and enjoyed doing it, not because it would further my career."[27]

Mary Ellen was able to integrate these different roles of her life smoothly, which enabled her to continue to be productive as a research mathematician when most other women of her generation did not. But as she points out in an interview published in *More Mathematical People,* the situation for young women today is quite different: they see themselves as professionals.

> If you do something professionally, it's harder for you to quit. You stay with it in hard times. The young women today are much more professional. We were amateurs. . . . We did what we did because we loved it. And some of us were lucky. For us, things worked out, and we did very well. Some were not so lucky, and they just dropped out along the way. But the young women mathematicians today are thinking in terms of a career from the beginning. It's true that they want a fulltime job. They want to do all the research in the world. They want to have a husband and children. They want to have a home. We wanted everything too—in spades—but the one thing we didn't demand, in fact it never occurred to us, was a career. The fact that they are thinking in terms of a career means that when it's a question of washing the socks or doing mathematics, they will often do mathematics. I think it was easier to quit doing mathematics in our day.[28]

Given Mary Ellen's vision of herself as a mother and housewife, how did she find her way into mathematics? And what kept her going?

Early Training

Mary Ellen Rudin had a rather unusual childhood and early training in mathematics. In both cases she was quite isolated from mainstream culture. She grew up in a little town in southwest Texas. It was a very isolated community, at an altitude of about 3,000 feet, tucked in a canyon formed by the Frio River and mountains on all sides.

In those days you entered the town by going fifty miles up a dirt road—you had to ford the river seven times to get there. . . . My father was there to build a new road, but the Depression hit and the State Highway Department never completed the road while we lived there. . . . It was a real mountain community. Many kids came to school on horseback. We had a well and an electric pump, which gave us running water, but most people didn't have running water or any of the things you think of as being perfectly standard. Yet it was wonderful. There were miles of wild country and beautiful trees along the river. Everywhere there was a beautiful view.

We had a very good school, and there were some very bright kids. We also had a lot of time to develop games. We had few toys. There was no movie house in town. We listened to things on the radio. That was our only contact with the outside world. But our games were very elaborate and purely in the imagination. I think actually that that is something that contributes to making a mathematician—having some time to think and being in the habit of imagining all sorts of complicated things.[29]

Growing up, she was not interested in mathematics (or any one particular thing) more than anything else; she certainly had no idea that she would become a mathematician. When she began college at the University of Texas, Austin, therefore, she simply followed her parents' guidance and decided to pursue a general education; in her first term she took a broad range of courses. However, as fate would have it, on her first day at registration she wandered over to the mathematics table, where there were very few people, and began talking with "an old white-haired gentleman" who asked her a lot of questions involving the use of terms such as *if* and *then*, *and* and *or*, to see if she used them correctly from a mathematical point of view. As it turned out, the man she spoke with was R. L. Moore, a mathematician famous for his unconventional teaching style, and for the number of mathematicians he produced. That conversation was to shape the course of her life. When she walked into her calculus class the next day, it was Moore who was teaching it. She went on to take many more courses with him, and she ultimately majored in mathematics. In fact, Moore was the only mathematics professor she studied with until her senior year. However, even then she still had no idea what she would do with her mathematics degree—being a high school teacher was the only thing she could imagine, and it was not what she wanted to do. But Moore had other plans in mind. She was invited to stay at the University of Texas for graduate school. She accepted the offer, and R. L. Moore became her graduate advisor.

I'm a mathematician because Moore caught me and demanded that I become a mathematician. He schooled me and pushed me at just the right

rate. He always looked for people who had not been influenced by other
mathematical experiences, and he caught me before I had been subjected to
influence of any kind. I was pure, unadulterated. He almost never got
anybody like that.[30]

Moore's methods were unusual and controversial. He had a very competitive and adversarial style, which was simultaneously meant to weed
out the weaker members and boost the confidence of the survivors.

> I was always conscious of being maneuvered by him. I hated being
> maneuvered. But part of his technique of teaching was to build your ability
> to withstand pressure from outside—pressure to give up mathematical
> research, pressure to change mathematical fields, pressure to achieve non-
> mathematical goals. So he maneuvered you in order to build your ego. He
> built your confidence that you could do *anything*. No matter what math-
> ematical problem you were faced with, you could do it. I have that total
> confidence to this day.[31]

In addition to developing their confidence, Moore taught his students
to be highly self-sufficient. In class, he would simply put definitions on the
board and give students potential theorems to work on. Some of the
theorems were true, others false, but the students had to discover which
were which on their own. They were discouraged from reading journals
or consulting other people; they were expected to work completely
independently. In fact, Moore had his own unique mathematical language, which made it even more difficult for his students to learn from
outside sources. He wanted them to focus on their own ideas, their own
insights, their own intuitions—in a sense creating mathematics from
scratch. At the time Mary Ellen wrote her thesis, she had never seen a
single mathematics paper.

Though the kind of independence they developed was a significant
asset, there were certainly problems with his methods as well. Many
students were turned off by the extreme degree of competition. Others
grew frustrated by the fact that their mathematical training made them
so isolated from what was going on in the field more generally. In fact,
Mary Ellen does not use his methods in her own classes. When asked how
she felt later about her mathematical education, she said:

> I really resented it, I admit. I felt cheated because, although I had a Ph.D.,
> I had never really been to graduate school. I hadn't learned any of the
> things that people ordinarily learn when they go to graduate school. I didn't
> know any algebra, literally none. I didn't know any topology. I didn't know
> any analysis—I didn't even know what an analytic function was. I had my
> confidence built, and my confidence was plenty strong. But when my

students get their Ph.D.'s, they know everything I can get them to learn about what's been done. Of course, they're not always as confident as I was.

Although Moore's primary goal was to instill confidence and independence in his students, what developed in the process was a very strong community among his students, which proved to be tremendously helpful in their careers and mathematical development. "We have all been very close to each other for our entire careers. That is, we were a team. We were a team against Moore and we were a team against each other, but at the same time we were a team for each other. It was a very close family type of relationship."[32]

This mathematical family played an important role in practical ways: helping her to get jobs, grants, and recognition. Indeed, when Mary Ellen finished graduate school, she never applied for a job; she was simply informed by Moore that she had a position at Duke, and it was assumed that she would take it. "Moore simply told me that I'd be going to Duke the next year. He and J.M. Thomas, who was a professor at Duke, had been on a train trip together. Duke had a women's college which was sort of pressuring them to hire a woman mathematician. So Moore told Thomas, 'I've got the very best, and I'll ship her to you next September.'"[33] He did, and Mary Ellen went.

In a similar vein, near the end of her position at Duke, one of her fellow Moore students, R. L. Wilder, arranged for her to go to the University of Michigan and even applied for a grant for her. He was impressed with her mathematical ability and was hoping to work with her there. However, Mary Ellen decided at the last moment to do something different, namely to marry Walter Rudin, who was then at the University of Rochester. In accord with the traditions of the time, Mary Ellen never questioned that she would follow her husband. Wilder, therefore, simply arranged for her grant to be transferred to the University of Rochester.

Marriage, Motherhood, and Mathematics

Over the course of the next decade, Mary Ellen Rudin had four children. Two of them came in the first two years after she got married. "I was very busy and very happy as a mother, and certainly expected to be a mother, and wanted to be a mother very much." It would not have been unusual to stop doing mathematics at that point—raising four children was a full-time occupation, and there was no incentive to have a career outside of the home. What, then, enabled her to continue on this path, even when many other even very gifted women did not?

To some extent her early mathematical training contributed to this commitment, not only by helping her to develop a highly autonomous work style and providing a community which propelled her along a mathematical path, but also through the underlying messages conveyed by R. L. Moore about what constitutes success and failure.

Moore had warned Mary Ellen about his earlier women students. One wrote a "fantastic thesis" but then immediately went off to China as a missionary. The other was "very influential as a teacher and administrator," but did not continue in research. Mary Ellen says that Moore "viewed his two earlier women students as failures and he didn't hesitate to tell me about them in great detail so I would realize that he didn't want to have another failure with a woman." In the context of other conversations he would subtly work on her—steering her to a life of mathematics. On one occasion when she mentioned something about having a husband someday, he responded, "Husband! But, Miss Estill, I thought that you were going to be a mathematician." But as Mary Ellen says, "Although he may have had his doubts, I never saw any contradiction in being both a housewife and a mathematician—of the two I was more driven to be a housewife."[34]

But other factors also played an important role in her continued commitment to mathematics. Like Fan Chung, Mary Ellen had significant help in raising her children. The Rudins' nanny, Lila Hilgendorf, was in many ways a second mother to the children:

> She cared for the children and did some housecleaning. When I would walk into the house, she would walk out; and when I had to go, she was there. We had that relationship until she died this year. She was absolutely the best mother I ever saw, and my children just adored her. One thing that she did for me was absolutely fantastic. The first weekend after my retarded child was born, she said, "oh, I just have to have him for the weekend!" And he went to her house for the weekend for the rest of her life. Actually he lived at her house the last several years. So when people ask me how it is, in my position to have four children, I have to say that when you've got Lila, it's easy. I am afraid that few women will ever have such an easy, non-pressured career as mine.[35]

Of course, central to this arrangement was the fact that they had the economic resources to hire such a nanny. They were able to live comfortably on Walter's salary, even with domestic help. And Mary Ellen did not need to work to make ends meet; hence she could devote her free time to research.

But equally important was the fact that both Mary Ellen and her husband felt comfortable hiring domestic help. As Mary Ellen points out,

"I think there are a fair number of women in my day who had the potential to be mathematicians, but who didn't become mathematicians because it didn't seem right to have someone help you with your children." Caring for the children was, after all, seen as a wife's primary responsibility. And even if a woman felt comfortable with such arrangements, often her husband did not. Walter, however, both understood and supported Mary Ellen's desire to do research. Hiring help seemed like a natural solution.

In Ravenna Helson's study of creative women mathematicians, there were very few things that the different women she interviewed had in common. One unexpected finding, however, was that most of the women she studied were married to Central European Jews. Many of these men, like Walter, had had nannies when they were young. Hence domestic help seemed normal; they recognized that "the person who changes your diapers doesn't necessarily determine what you do in life. It doesn't determine one's values." Furthermore, they came from a cultural milieu that celebrated intellectual wives—having a wife who was a mathematician was therefore seen as an asset, not a problem.

Mary Ellen and Walter's marriage was a significant factor in enabling her to stay actively engaged in mathematics in other ways as well. Because Walter is a highly respected mathematician, it has been fairly easy for Mary Ellen to integrate into the mathematical community. Both the University of Rochester and the University of Wisconsin were glad to have her teach; she could use libraries and other resources, and she was given an office and invited to seminars. Certainly Mary Ellen's talent and her ties to the Moore family helped, but if she had had no direct connection to these universities, it is unlikely that they would have given her such a warm reception. This kind of community support has been quite important in sustaining her mathematically.

> The whole community support—family and other mathematicians' support—are needed in order to be a mathematician. You may be able to get along without some of it, up to a point, but it certainly makes it more probable that you will be an effective mathematician if you have it. It allows you to focus on your mathematics problems. If you look at the mathematicians from the nineteenth century, like Grace Chisolm Young, Emmy Noether, Sofia Kovalevskaia, they were people who had to completely upset the community around them. They were strong enough to do it. But I'm sure there were infinitely many more potential mathematicians who would have been glad to pursue mathematics if they had the opportunity to go to school, and social acceptance for what they were doing. It was hard; I think none of them were really happy women in some sense. Each of them had a certain kind of support, which perhaps was all they cared about. But

their community of support was in the middle of a hostile environment. It is helpful if the environment isn't hostile.

Being married to a mathematician had other benefits as well. The Rudins could travel around the world together to attend conferences, or give talks, and these trips were invariably both mathematically stimulating and exciting vacations. It was not unusual for one of them to give a lecture in the morning, and then be out on the beach in the afternoon. The walls of their house are covered with pictures of the their trips to places such as Hawaii, New Zealand, and China. They think of their colleagues, friends, and students as their mathematical family—one that extends across the globe.

Although being married to Walter helped create the kind of supportive environment that Mary Ellen points out is quite helpful, she and Walter do not actually do mathematics together, or even talk with each other about their results. They are in different fields, and have different work styles. Nor do they feel the *need* to share that part of themselves with each other. "Everybody should have something [of their own]. I have plenty of things to share with people. I don't have to share all of my mathematics with them, too." In addition, she and Walter have very different ways of thinking about mathematics and working on problems. Mary Ellen describes herself as a problem solver rather than a theory builder; it is a style that lends itself to her field of topology. Walter, on the other hand, is more concerned with building structures, a skill that is particularly useful in his field of analysis.

Moreover, Mary Ellen's way of doing mathematics is highly individualistic. She likes working alone because she thinks quite differently than most other mathematicians she knows—she is very visually oriented. As she says, she cannot follow a proof or an argument without paper and pencil in hand; she relies on pictures to help visualize an idea. "I don't learn mathematics with my ears." As a result, there are very few colleagues or students with whom she can sit down and do mathematics spontaneously. More typically, a student will come in with a problem, they will discuss it briefly, she might make a suggestion or two, but then they go off and work on it separately. The next time they come together, she will have a few more suggestions. Both her natural style and her early training contribute to her preference for working alone.

Ironically, while Mary Ellen is accustomed to working alone in one sense, she is often far from alone in another. It is not unusual for her to work at home—not in a study with her door closed, but rather sitting on her couch in the middle of the living room, surrounded by children.[36]

> It's a very easy house to work in. It has a living room two stories high, and everything else sort of opens into that. It actually suits the way I've always

handled the household. I have never minded doing mathematics lying on the sofa in the middle of the living room with the children climbing all over me. I like to know, even when I am working on mathematics, what is going on. I like to be in the center of things, so the house lends itself perfectly to my mathematics. . . . I feel more comfortable and confident when I'm in the middle of things, and to do mathematics you have to feel comfortable and confident.

For some people, the distractions of home make it impossible to do creative mathematical work. But Mary Ellen has an amazing ability to work even in the midst of domestic activity. Part of what enables her to do this is the grit and determination that are a natural part of her work style. She is the kind of person who simply cannot let a problem go once it has taken hold of her. She describes what she does when she gets stuck on a problem:

I am apt to hit it directly on the head, again and again and again and again. I find it very difficult to give it up. I don't try to find an easier problem elsewhere. I know mathematicians who do that. Some people are very good at going sideways, or going for a partial solution, but I never want to turn aside. I want to go forward and go for the gold. I struggle with myself to put a problem away, when I have looked at it a thousand times and drawn the same picture a thousand times and gotten no idea. But I find it very painful to put something away. I can't let go.

This kind of intense determination and focus is extremely useful in maintaining the drive to do research even through the inevitable struggles and frustration that arise in working on mathematics.

Clearly, then, there were many factors that contributed to Mary Ellen's ability to continue as an active research mathematician, even as she raised four children and created a home life. In the end, however, Mary Ellen argues that while all the external factors such as economics, domestic help, and community and family support are important, they alone will not sustain you. You should do mathematics only "if you really want to, and only if you enjoy working hard on hard problems, and only if you find it tremendously satisfying to solve difficult questions."

Is This a Viable Model Today?

Some would argue that while Mary Ellen Rudin's situation was in many ways ideal, it is no longer a viable option for most women entering mathematics today. First, few families can survive on one person's income. The model of the husband as breadwinner and wife as homemaker is no longer an economic possibility for many families. Mary Ellen's arrangement is even less an option for women who are single or self-supporting.

Second, many women find it difficult to be accepted by the mathematics community if they are not seen as professional. While the label "amateur" mathematician may have been an honorable one when women were not expected to do anything other than domestic work, it is no longer a label that would meet with such acceptance. Such a designation would now have negative connotations, and would make it difficult, for example, to receive grants, awards, jobs, and recognition.

Third, it is not only how women are perceived by their colleagues that is at stake, it is also a matter of how women perceive themselves in relation to their work. Women are raised with the expectation that they will be just like men in the commitment to their work—they expect to have equal opportunities and equal treatment. Hence, many women would feel uncomfortable defining themselves as amateurs. They see themselves, and expect to be seen, as professionals.

Nonetheless, Mary Ellen's model is an important one, and should not be dismissed too quickly as no longer relevant. It is an example of a woman who was able to successfully combine marriage, motherhood, and mathematics. She was able to stay active and productive as a researcher. She had four children and was actively involved in raising them, and she did not feel that she had to deny the domestic side of her life. Perhaps most important, she was very happy.

Is it possible, therefore, to modify her model to meet the needs of contemporary women, and increasingly of men as well? Is it not possible, for example, for mathematicians to be part-time for certain segments of their career, and still be defined as professionals? Is it necessary to maintain such a degree of separation between work and family in order to be taken seriously as a mathematician? Is it necessary to relegate family to a lesser priority in order to be defined as a real mathematician? Certainly Mary Ellen Rudin suggests a different way of looking at these questions. She would argue that maintaining some connection to research at all times, even when children are young, is very important, but as her life illustrates, there are many ways to do this.

Fan Chung

Fan Chung

(1949–)

Fan Chung is the only woman interviewed for this book who was born in another country—she grew up in Taiwan and received her early mathematical training at Taiwan University. Chung came to the United States to do graduate work at the University of Pennsylvania and continued as a research mathematician in industry, first at Bell Labs, and later at Bellcore. Most recently she became a professor of both mathematics and computer science at the University of Pennsylvania. Chung's training in Taiwan, as well as her work in industry, provides an interesting contrast to the other stories included here.[37]

Though it is rather unusual for women to be mathematicians in either China or Taiwan, Fan's class in mathematics at Taiwan University was one-third female. The camaraderie with her female classmates, most of whom continued on in mathematics in the United States, played an important role in her identity as a woman mathematician. Fan has been an extremely productive mathematician; she has published more than 170 papers, with more than 90 different co-authors. For Fan, collaboration is a powerful mode of doing mathematics, one that she actively cultivates not only in her own work, but also in her role as a manager of a team of researchers at Bellcore.

Fan's life as a woman in mathematics has been a fairly smooth one. She was able to integrate marriage and motherhood with being a mathematician. Like so many women mathematicians, she is married to a mathematician, Ron Graham. Their relationship exemplifies one way to integrate personal and professional life and illustrates some of the advantages as well as potential disadvantages of such an arrangement.

From Taiwan to America

Fan Chung's family was from mainland China, but they moved to Taiwan the year Fan was born. Fan stayed in Taiwan until she finished college, at which point she came to the United States to begin graduate studies in mathematics. Her family was unusual because her mother worked full-time; she was a high school teacher who was respected and loved by her students. Though her mother taught what sounds like a very traditional subject for women—home economics—in Taiwan she represented a modern and non-traditional woman, both in her lifestyle (having a full-time career) and in what she taught. Home economics signified not so much the traditions of the past as the technology of the future, including the use of sewing machines and kitchen appliances, which radically redefined domestic roles. Fan's mother passed on to her daughter the lessons and advantages of economic independence, strongly encouraging her to pursue a career and "not just to be an attachment to a man."

It was clear to Fan, however, that she could not follow in her mother's footsteps by pursuing home economics. Though she much admired some of the craft traditions, including the very old art of knot design, she was not particularly good with her hands, so she decided early on to pursue a life of the mind instead. By the time she was in high school, she had decided on mathematics. "My father sort of directed me to mathematics. He said mathematics is a foundation and you can always switch to other fields later on. Also, since I'm not so good with my hands, it seemed good to go into a science that didn't require experiments, just a pencil and paper." Because Fan excelled on the standardized aptitude tests that were given to all students, her path into mathematics was relatively easy. She was the top math student in her high school on these tests, and was admitted to one of the best colleges, Taiwan University. Fan is humble about this accomplishment, for she believes it is really "small cleverness" to do well on exams. The real challenges in becoming a mathematician were yet to come.

As an undergraduate at Taiwan University, Fan received an extremely rigorous education. After her first year, *all* of her courses were in mathematics, a degree of specialization almost unheard of in American undergraduate education. The courses were intensive, many equivalent to graduate-level courses in the United States. Fan enjoyed this immersion in mathematics. Her positive experience was partly due to the unusual demographics of her class—out of thirty students, ten were women, a strikingly high percentage for Taiwan University. And the women were by far the strongest students; each had been either first or second in her high school. This community of women created a supportive environment within which to learn mathematics. "When we were all in the same class, we would help each other a lot. It's not competition at all." As Fan elaborates:

> I was quite fortunate because it's good to be able to have mathematical discussions with other people. It's an important part of education. But in Taiwan, it's quite conservative; there were only male professors, no female professors. And the boys mingled together socially, but the girls were out! Girls were not in their circle at all. Fortunately there were several of us so we could discuss things among ourselves. In general women graduate students are at a disadvantage. They cannot mingle as much because it's a social situation.

Many of the women in her class went on to do graduate work in the U.S., and to have successful careers in mathematics. Though they don't get to see each other a lot, they still stay in touch and know what is going on in each other's lives.

Having support in making the transition to the United States was important because it was quite a difficult change. For some, the culture shock was unbearable. The smartest of the group, Fan's closest friend, ended up committing suicide. She was doing fine mathematically, but the foreignness was extreme: the language, the customs, the values. It is impossible to know what actually led to the tragedy, or whether the same thing would have happened had she stayed in Taiwan. As Fan says, "There is often a very thin line between being talented and being crazy."

One way Fan was able to cope with the transition to America was to marry early. In this way she was protected from having to deal with the most radical differences in culture, particularly in the social sphere:

> For us, holding hands with a boyfriend was a serious thing. But the first thing I learned when I came to this country was that couples who were not even married were living together. According to my teachings, that was absolutely a no-no. It was a difficult change. My solution was to get married, so I wouldn't have to deal with all that confusion.

Graduate School and Early Professional Life

Fan went to graduate school at the University of Pennsylvania in Philadelphia. In contrast to undergraduate school, where she did most of her work with her fellow students, in graduate school she worked primarily alone or with her advisor, Herbert Wilf.[38] There was one other female graduate student, Joanne Hutchinson, whom she really liked and who would talk with her about both mathematics and social issues. The latter was particularly helpful for Fan.

From the start, it was not difficult for Fan to integrate mathematics and motherhood. She had her first child during her last year of graduate school. She believes that this is in fact a good time for women to have children. In Fan's case, it was made easier by the fact that her schedule was quite flexible. She had done so well during her first years as a graduate student that she received a research fellowship, which precluded the need to teach. Hence, she was able to work at home whenever she needed to, both throughout her pregnancy and during the first few months after her child was born. The only problematic part was that she was also applying for jobs, and she thought it would hurt her professionally to go to interviews so obviously pregnant. The job market was tight, she was a woman, and she was from a foreign country to boot. Being pregnant did not seem like an asset. Fortunately, she received two job offers without an interview. One was a short-term position at the University of Hawaii. The other was a position at Bell Labs, a major center for mathematical research. She chose Bell Labs. She had done a joint paper with Ron Graham, a highly respected mathematician there. Based on that work and word-of-mouth recommendations, she was hired.

Fan's research has been in combinatorics and discrete mathematics—areas that could have potentially important applications in the communications industry and beyond. Combinatorics is an unusual field; the problems are often simply stated, and in that way very accessible, though they can still taken tremendous creativity and ingenuity to solve.

> Combinatorics was really interesting and fun for me. Some people with infinite training in mathematics, or too much theoretical buildup, sometimes have a hard time dealing with these simply stated problems. They don't know where to start. There's no clear, well-defined path. You have to just look at it. Of course, that's not completely true. There are some standard kinds of techniques. And there are some well-developed theories, but often you have to use everything you know, and just dig in with your hands. I like this problem-solving kind of thing. There are two kinds of mathematicians, theory developers and problem solvers, I think I have a good part of the problem solver in me.

Fan stayed at Bell Labs from 1974 to 1983, until the AT&T antitrust lawsuit. After the breakup, an offshoot, Bellcore, was formed, where Fan is now a division manager and continues to be actively involved in research.

Being active in research did not, however, preclude her having a second child. When she got pregnant during her second year at Bell Labs, her manager wondered whether she planned to quit once the child was born. Fan never discussed her home life with her professional colleagues, so her manager was not even aware that she already had one child, and that that had clearly not interfered with her work.

> I just said that I would work until the last day. It's a lot of trouble in this country to take maternity leave. They have to change benefit plans, etc. So I said I would just take vacation. It was not even in my record that I was leaving for a second child. In fact, I actually wrote a paper during the four weeks I was out. I was in the hospital and then home, and Ron mentioned a nice problem to me—a graph decomposition problem. I heard the problem and immediately thought I could do something about it. So I solved the problem and wrote up a paper during that month.

Once again, Fan found having a child relatively easy to integrate into her work life. She had a private office, so she could close the door and take a nap when she needed to. Her schedule was again quite flexible—she did not have students knocking at her door or courses to teach. She could work any time, day or night. But perhaps most important, she had tremendous support at home. She was able to hire excellent women to help with the children and domestic responsibilities.

> Having live-in help made a big difference. It takes away the worrying. Some women enjoy staying at home, and that's fine. But for those who like to have their own work, they have to get as much help as they can—from husbands, or someone else—and try not to put too much on themselves. I have seen some women try to do too much themselves, and they get too tired. It's not good for them or the kids.

Her husband was also quite helpful with their children, a somewhat unusual trait for a man from Taiwan. For some women, hiring full-time help to take over duties traditionally ascribed to women is a difficult step. They feel that they are abandoning their duties as mothers, and are inclined to try to do it all themselves. But this seems not to be an issue that Fan struggled with. In part, this was because she had clear signals from both her parents, even when she was young, that having a professional life is very important.

Although having hired help was essential, it did not eliminate all domestic responsibilities. One final factor she feels is critical for balancing personal and professional life is the ability to juggle these different roles simultaneously.

You have to be able to juggle. When I was at home, I was cooking, thinking, working on my stuff, while the kids kept coming by and asking questions. You have to piece together small periods of time and accept interruptions. House things need to be done, kids need attention; husbands are almost like kids, they need attention too. If you really need uninterrupted, concentrated time, it's hard.

Thus, while for Fan integrating mathematics and motherhood was relatively easy, all of these factors were important: talent, a flexible schedule, smooth pregnancies, healthy children, excellent live-in help, a supportive spouse and parents, and the ability to juggle many things at once.

A Mathematical Marriage

Just as Fan was able to mix mathematics and motherhood, so too was she able to smoothly integrate marriage and professional life. While Fan's first marriage played an important role in her adjustment to the new world of the United States, and her husband was actively involved in raising the children, in the end it did not last. "The marriage was not so bad, but I think I went into it a little bit too fast. We had a few good years, but it was not the right foundation for a marriage. He was very nice to me, but I was never madly in love."

Her second marriage, to Ron Graham, a talented and eclectic mathematician who also worked at Bell Labs, is a striking example of how personal and professional life can become completely fused. Though many women mathematicians are married to other mathematicians, few have such an intensely collaborative relationship. As Fan says, "In terms of mathematics, we have almost our own language." Mathematics, to them, is not so much about work as it is about play. They love what they do; it is fun and interesting, and exactly what they want to be doing with their time—all of their time. As Fan says, they are both workaholics. It is not unusual for mathematics to be the first thing they talk about when they wake up in the morning. Even in the evening, when they arrive home from work, they do mathematics together. And on the rare occasion that they take a vacation, they do mathematics wherever they go.

Their house, too, is a perfect reflection of the bond that unites them. It is a lovely and spacious home, recently remodeled with an addition that Fan designed. One room is their mathematics library, filled with books and mathematics journals from floor to ceiling. Their living room has one

whole wall dedicated to their whiteboard—the kind with markers so they
don't have chalk dust all over the room. Even their furniture was carefully
chosen to add to the functional ambiance: a beanbag chair, a hanging
chair, and other unusual chairs.

Their style of doing mathematics is casual and playful. They don't sit at
a desk or a computer. "I usually work by holding a pad and pencil and sit
in one of those chairs. In our bedroom there is a round glass table. Now
and then I sit there and do things, but it's not like a desk; there are no
drawers, it's just a clear surface." There are plenty of diversions around:
Ron is an avid juggler and trampolinist, there are rooms full of video
arcade games, and puzzles, Rubik's cubes, and electronic teasers are
scattered throughout the house.

For many people, close collaboration is a difficult relationship to
achieve. How do they manage, given how intimately intertwined their
lives are? Fan acknowledges that she has had to learn how to work with
Ron, since their styles are somewhat different. Occasionally she feels
strongly and differently about something, for example how a paper should
be written up, but for the most part their collaboration is easy. She has
learned how to work with their differences:

> Ron is a very smart man, so it takes a relatively smart person to
> understand him. He definitely has many ideas, ranging from new and
> interesting to outrageous. Sometimes if you push him to the north, he goes
> to the south, so if I want to get him to go to the south, I push to the north!
> After understanding that theory a little bit, I found that he is really very
> easygoing and considerate.

> *You know him well.*

> Yes, actually we even joke often about that. That's probably part of the
> reason that some couples don't work together—it could cause too much
> friction. But in our case we really enjoy working together a lot. I do notice
> that the work gets much better, not just with Ron but also with other co-
> authors. With another person pushing in a different direction it goes faster
> and further.

> *Two halves make more than a whole.*

> Right.

Collaboration occasionally extends to teaching as well as research. She
and Ron recently co-taught a course at Princeton based on a book Ron co-
authored called *Concrete Mathematics*.[39]

Being married to Ron has other advantages as well. His reputation gives
him the flexibility to accommodate Fan's career. At one point she was

offered, and seriously considered taking, a position at a major research university. Ron was very supportive since it would have been relatively easy for him to find a job in the new location. Though she declined the offer in the end, such options continue to be a possibility in their lives.

Mathematics is a passion they can share not only with each other, but with their children as well. At the time of this interview, Fan's son was at a math training camp where future math Olympiad team members are selected. Both Fan and Ron had spent weeks coaching him, and Ron had developed a wonderful series of lessons to teach him mathematics. Though Fan felt that her son's chances of being selected for the team were slim because this was his first year at the camp, the whole family was very involved in the excitement and training.

Of course, there are certain disadvantages in being married to a mathematician as well. Fan is keenly aware of the potential danger of living in the shadow of such a prominent mathematician. She has been careful, therefore, to do plenty of research on her own and with other mathematicians. Her work with Ron is only about one-third of her published research.

Another disadvantage arose when Fan was considered for a political office in a major mathematics association. She was turned down because she was married to Ron, who had a formal position in the association. It was seen as a conflict of interest for her to have such a post. She was somewhat disappointed. "In that sense it hurts me to be married to such a big name. It hurts my identity. As a woman mathematician, it's relatively hard not to marry another mathematician, just because most people I run into are also mathematicians. We share more in common and we work on problems together. It's wonderful to have a spouse who can appreciate what's been done and who can say, 'Oh, this is nice.'" It is rare that a man is denied a position because his wife is a member of the same professional organization.

Mathematical Research

One reason Fan has so thoroughly enjoyed her field of research—combinatorics—is that it has connections to many other fields within mathematics, as well as to other sciences and technology. These interdisciplinary connections contribute to the rapid growth of the field, continuously stimulating new ideas and problems. At times she works with engineers, at other times with computer scientists, and at still other times with number theorists. And because the field is quite young, there are plenty of problems to work on. In fact, as she says, "there are too many problems and not enough ideas. I can just stretch my arm in the hole and pick one out."[40]

In the beginning she would work on any problem she came across that looked interesting and fun, but over time she has come to have a larger vision of combinatorics research, and she now tries to work on problems that seem important or central. She has developed a "taste" for problems as she sees how things fit together into "sort of a bigger tree." The question of what constitutes an important problem is of central concern to her in her role as editor of a major mathematical journal. And it is a topic she discussed recently on a trip to China, for she feels that one of the major difficulties for the Chinese is that they are relatively isolated mathematically and hence cannot always discriminate between dead-end problems and those that are more central.

So what criteria does Fan use in determining important questions? First, she argues that it is good to work on "mainstream" problems, i.e., central problems in a field that may relate to many other problems. When you work on mainstream problems, even partial results can be important. Furthermore, even if you don't solve them, you will learn a lot of important mathematics by working on them. Second, she recommends working on problems that other (prominent) people have asked, and worked on, but have not been able to solve. The underlying assumption is that major researchers in the field have developed a taste for important problems, and so one can rely on their judgment. A third guide is to find problems that have a lot of impact either within the field of combinatorics or on other fields in math (pure or applied) and science. Often the impact of a problem is not recognized until after it has been solved, however, so this is only partially helpful. Finally, Fan says, it certainly helps if a problem has "a natural beauty."

Fan recognizes that even though all of these factors are important, often people choose to work on a problem because it captures their imagination and stimulates their curiosity. A "natural curiosity" is something she refers to often and believes is very important for a mathematician.

One of the advantages of working in industry is the tremendous amount of collaboration among colleagues. Communication between mathematicians is actively cultivated, and there is a great deal of flexibility in schedules to allow for travel and interaction with other mathematicians around the world. "When you think about it, all you need [for research] is a computer, travel money, help typing up papers, etc. But that's about all." And a library as well. But what is clear is that communication with colleagues is at the core of research. And research labs such as Bell Labs and Bellcore have done a good job at creating that. Indeed, they provide a fascinating contrast to the model that has evolved in academia, which tends to be focused more on the individual. The whole system of tenure and promotion tends to cultivate competition rather than collaboration. Clearly, however, research in industry is effective—most labs would not survive if they did not prove to be worthwhile. Though research in this context is inevitably more closely tied to utilitarian ends, there is no reason to assume that a similar model might not be effective in research that was not driven by such clear practical goals.

Thus, Fan's story raises provocative questions about how mathematics is practiced. Certainly for Fan, collaboration has been central to her research. As she says, "My co-authors are my best teachers. You learn a lot of proven theorems, known results, and how to actually use them to do things. You really see the action when you collaborate with other people. . . . It's a wonderful relationship—it's a little more than just friends."

She finds the four axioms of collaboration described by Hardy and Littlewood to be helpful guidelines for successful collaboration:

1. You don't have to be right or complete in what you say while collaborating.

2. When one person tries to communicate (e.g., by letter), the other has no obligation to respond because people have different paces. This is a way of protecting each other's freedom.

3. It is not necessary for both people to check all the details.

4. The proportion of contribution is totally irrelevant.

Fan thinks the last axiom is the most important, for she believes that the cause (i.e., getting results) and the relationship are much more important than quibbling about who contributed how much. Inevitably things balance out in the end, and the relationship is more important than a single paper because it often leads to future collaboration and results.

In moving from Bell Labs to Bellcore, Fan moved into administration. She is in charge of the research division in mathematics and theoretical computer science. In this capacity, one of her primary tasks is to create an atmosphere conducive to creativity, productivity, and collaboration. She is careful not to intimidate or take advantage of young researchers in her division. Thus she is sometimes less likely to do joint work with them. Instead, she encourages them to work with other peers, so they won't get lost behind her name.

Fan also believes it is important to recognize that different people have different mathematical paces: both in the short-term sense—how quickly they can solve a given problem—and also in the long-term sense—at what point in their life they do their best work. In her own case, she feels fortunate to be fairly fast at doing math.

> I have had a chance to meet and work with many great mathematicians. Some people are just very, very fast. And that's good too. I've been able to work with them and you get a little faster. I think I'm faster than most people, but there are a few people who are just so much faster than I am. On the other hand, I know some great mathematicians who are very slow. So slow probably helps because it helps you think differently. I certainly admire those very fast people. That's one way of showing their intelligence. Some people are so fast they start to intimidate other people, and I think that's not very nice. I've been on both sides. There are people who are slower, especially young people. I always try to tell them that the speed is not the thing, it's the quality of research that matters. These two things are not always so perfectly correlated.

Fan is grateful that she ran into people who were faster than she late in life, for by then she had developed enough confidence in her own abilities that she was not easily intimidated. Instead, she just tries to learn from such colleagues and appreciate their speed for what it is.

Pace enters in not only in doing a given problem, but also in the timing of one's life and career. While there is the prevalent myth in math that one does one's best work early in life, Fan has found quite the opposite to be true. As she says, "Actually, I think I've proved my better theorems recently. I think my last year's work was my best. People have different paces."

Certainly being married to another mathematician and being in a highly stimulating research environment both contribute to her continued productivity. But in the end, what is perhaps most sustaining of all is that she continues to find mathematics both interesting and fun. "I think if we want to encourage more people into research, they should know the existence of such a great life. . . . Mathematics, especially research in mathematics, is wonderful. It's big puzzles and small puzzles. It's more than solving a puzzle. Like Hardy said, it's uncovering structures. In many senses it's like art work."

Epilogue
by Fan Chung

About five years ago, I was awarded a Bellcore Fellow and took a sabbatical as a visiting professor at Harvard. At the time, my choice was between a place with many people in my area or Harvard, where there was no one in combinatorics. I chose the latter, and it has worked out well. I ran into Shlomo Sternberg in the first week, and we started a series of collaborative works, including one article on the mathematics of the Buckyball which appeared in *American Scientist*. I have also collaborated with David Mumford and Persi Diaconis, repectively. Perhaps the most important collaboration is with S. T. Yau. We have worked on a series of ideas to develop spectral graph theory using powerful techniques in spectral geometry. The interplay between the discrete and the continuous has brought new approaches, deep insight, and many results in both areas. The work is still continuing since many new directions are being opened up. In the article, I talked about problem-solving and theory-developing. I guess that there is quite a bit of theory-developing in me also.

As you can see, the visit to Harvard was really good for my mathematics. At the time, the decision to leave our comfortable home to commute 250 miles away was not an easy one. It was possible only with the strong support of Ron, who understands and encourages the need for growth.

Two years ago I joined the faculty of the University of Pennsylvania as professor of mathematics and also the endowed Class of 1965 Professor. Recently, the Department of Computer Science at Penn gave me a secondary appointment as professor of computer science.

I have enjoyed the move to academic life. Some courses that I am teaching aim at bringing out the power and beauty of math by conveying the interaction of different mathematical areas and the connections of theory and practice. We have a number of guest speakers to address how mathematics is actually used as an integrated part of current technology in computing and in communication.

3

Is Mathematics
a Young Man's Game?

Myth: Mathematicians do their best work in their youth.

In doing research for this book, I was struck by the fact that almost every single one of the women interviewed felt that she had done her best work later in her life, typically in her forties, fifties, even sixties. This was surprising to me. I had always accepted the prevalent belief that mathematicians do their best work in their youth. As the eminent mathematician G. H. Hardy wrote in his influential book *A Mathematician's Apology,*

> If then I find myself writing, not mathematics but "about" mathematics, it is a confession of weakness, for which I may rightly be scorned or pitied by younger and more vigorous mathematicians. I write about mathematics because, like any other mathematician who has passed sixty, I have no longer the freshness of mind, the energy, or the patience to carry on effectively with my proper job. . . . No mathematician should ever allow himself to forget that mathematics, more than any other art or science, is a young man's game.[1]

While I had questioned that mathematics was a young *man's* game, it had not occurred to me to question that mathematics was a *young* man's game. Hardy is not alone in this assumption; many other mathematicians have expressed similar views. Alfred Adler, in his essay "Mathematics and Creativity," is even more explicit about the time of life identified with productivity:

> Such consuming commitment can rarely be continued into middle and old age, and mathematicians after a time do minor work. In addition,

mathematics is continually generating new concepts, which seem profound to the older men and must be painstakingly studied and learned. The young mathematicians absorb these concepts in their university studies and find them simple. What is agonizingly difficult for their teachers appears only natural to them. The students begin where the teachers have stopped; the teachers become scholarly observers.[2]

In a similar vein, Andre Weil writes: "Mathematical talent usually shows itself at an early age. . . . There are examples to show that in mathematics an old person can do useful work, even inspired work; but they are rare and each case fills us with wonder and admiration."[3]

And as Sylvia Wiegand recounts in an article about her grandmother, the mathematician Grace Chisolm Young: "On his fiftieth birthday, Klein was honored in Turin where Grace [Chisolm Young] was then studying. At dinner he was seated next to Grace, said to be his favorite pupil, and he whispered to her: 'Ah, I envy you. You are in the happy age of productivity. When everyone begins to speak well of you, you are on the downward road.'"[4]

The contrast between these words and the experiences of the women I interviewed led me to wonder: Is it true that mathematics is a young person's game? What does happen as mathematicians age? And how does age intersect with questions of gender?

Is Mathematics a "Young Man's Game"?

When we look at science more generally, the answer to this question is no, science is not a young man's game. Numerous studies have found no positive correlation between youth and productivity in science.[5]

But what about mathematics? In many experimental sciences, it could be argued, the need for expensive equipment, access to funding, and ties to research communities have a more powerful influence on productivity than they do in mathematics. Since these factors improve with age (as one becomes a more established scientist), these fields would be less prone to favoring youth. In mathematics, however, one can do research with virtually no funding or equipment. Moreover, mathematics is at least perceived to be solitary work, and for this reason one might assume that there is less dependence on the community. Is it true, then, that in mathematics, productivity *is* correlated with youth?

Despite the pervasiveness of this belief, few systematic studies have been done on this question. One useful exception is the work by Nancy

Stern, published in her article "Age and Achievement in Mathematics: A Case-Study in the Sociology of Science."[6] Stern studied the relationship between age and mathematical productivity as measured by both quantity and quality of research.

Quantity of research is fairly straightforward to measure—one can simply count the number of articles published by a mathematician.[7] Using this measure, do mathematicians indeed produce a greater quantity of research when they are young? Stern's findings match those of the work done on scientists more generally. She found no correlation between age and productivity:

Age and Mathematical Productivity, 1970–74:
Mean Number of Papers Published in 1970–74
by Mathematicians of Different Ages[8]

Age	Mean Number of Papers
under 35	5.12
35–39	7.33
40–44	6.24
45–49	3.49
50–59	5.22
60+	6.11

In this study, the mathematicians who were most productive were between the ages of thirty-five and thirty-nine, and the number of articles published by mathematicians over the age of sixty was still quite high, surpassing the number published by those under thirty-five! So much for the hypothesis that mathematicians are most productive before the age of thirty-five. One might first be tempted to argue that the papers published by those over sixty were primarily co-authored with younger and more vigorous mathematicians. But in fact, Stern's findings show that the average number of co-authored papers by mathematicians over the age of sixty is still relatively low, and the number of singly authored papers by those in this age group remains high.[9]

The second natural challenge, then, is to argue that while quantity may still be high late in one's career, *quality* may steadily decline over time. Quality of research is somewhat more difficult to assess. However, a standard method used to evaluate quality is to determine the number of citations a mathematician has received for published literature. The more citations, at least in theory, the more important the work. As Stern points out, there are problems with this measure, and some have argued that it does not necessarily capture the most important work in a field. For example, in some cases an important mathematical idea can "finish off" a field. That is, the most important questions get answered, and little more is left to be done. In such a case, few new mathematicians would pursue that particular area; over time the field itself dies out, and the number of future citations may be minimal, thereby masking the importance of the result.[10] Nonetheless, Stern gives compelling evidence to suggest that the number of citations *is* a reasonable measure of significance.[11]

Once again, Stern's findings were similar to the research done on scientists in general: there is no simple correlation between age and *quality* of research. Stern found that the number of citations for work done in a four-year period for mathematicians over the age of sixty was almost double that for mathematicians under the age of thirty-five: "In short, no clear-cut relationship exists between age and productivity, or between age and quality of work. The claim that younger mathematicians (whether for physiological or sociological reasons) are more apt to create important work is, then, unsubstantiated."[12]

What about Women?

These findings are also corroborated by the interviews in this book. Again and again, the women state that their best work was done later in their lives. As Fan Chung said, "I think I've proved my better theorems recently. I think my last year's work was my best. People have different paces, especially women. Some people peak later in their lives."[13]

Marian Pour-El says, "Perception feeds upon perception. The awards that are given are given to very young people. You're over the hill if you're in your late thirties. I've never felt that. That's not my personal perception of the way I do things. I think I've done my best work later on, by a long shot."

Joan Birman felt that she was better able to focus on math later in life, after the issues of marriage were settled, her children were older, etc.. As she says, "I think doing mathematics when you're enthusiastic is important—not your age."

Judy Roitman agrees. When asked whether she believes that mathematicians do their best work before the age of thirty or thirty-five, she replied,

> No, but what I do think is that people naturally get tired. It helps to make shifts in your career. If you do the same thing all the time, then you die. If I look at the people I know who are in their fifties and sixties who have essentially stopped doing a lot of research mathematics, I think boredom

sets in. I don't think it's the brain that stops working. I think interest is a very, very important thing.

Mary Ellen argues that one's mathematical environment plays a very important role in mathematical productivity. For example, the University of Wisconsin has been quite stimulating for many mathematicians, including Mary Ellen. It is a large enough institution that mini-communities can form around mathematical specialties, one is constantly meeting new mathematicians and hearing new ideas, and the environment is described by many as supportive and encouraging. (Judy Roitman attributes this in part to Mary Ellen's presence.) Mary Ellen says of the myth that mathematicians do their best work in their youth: "The University of Wisconsin is a counterexample. Very often people have more or less picked out the field they are going to work in by the time they are thirty years old, but I don't think most people's best work will be done by the time they're thirty, and certainly my best work wasn't done until I was fifty-five years old."

Impact of the Ideology of Youth

In spite of evidence to the contrary, the image of mathematics as a young man's game has a powerful hold on the imagination of the mathematics community, and can have a particularly detrimental effect on women. The ideology of youth affects practices and policies that shape graduate programs, hiring, and the granting of awards.

Graduate programs are typically geared toward a young, fresh-out-of-college population with no family commitments. Students are expected to be full-time, and to have few other social or economic obligations. This intense focused period is supposed to coincide with the budding of one's mathematical creativity. Similarly, most universities prefer to hire young mathematicians—the potential hot stars; youth is considered a sign of precocity and brightness. Once again this early period is seen as key to one's mathematical productivity.

Even some of the most prestigious awards, including the Field's Medal (the mathematical equivalent of the Nobel Prize), have an age limit on who can win them. Field's Medal winners must be no more than forty years of age. Given the findings of Stern's studies, it is clear that often a mathematician's greatest work would not fall under this stipulation.

These kinds of preferences in policies can be particularly problematic for women. For many reasons, women's professional timelines—what they accomplish when—may be markedly different from those of their male counterparts. Women are less likely to have a clear linear professional path, and their lives are more likely to be interrupted for a period of time, or to have the veering and tacking quality described by Aisenberg and Harrington in their book *Women of Academe*. Factors that affect women's timelines include having children, following spouses' careers, lack of confidence in mathematical ability and hence more difficulty envisioning themselves as mathematicians, less explicit advice and counseling on mathematical careers, less encouragement from teachers, peers, and advisors to pursue mathematics professionally, and more difficulty integrating into the social fabric of the mathematics community.

Each of these topics is complex, and they affect different women to varying degrees (and at different times in their lives). One must be careful in drawing conclusions. For example, to what extent, and in what way, does having children affect women's mathematical lives? Does it, for example, make them less productive in their youth? Certainly in some cases, having children did have a significant bearing on timing and productivity. Joan Birman, for example, did not get her Ph.D. until the age of forty because she was raising three children (as well as working part-time and following her husband in his career moves). Mary Ellen Rudin maintained part-time positions because they were well suited to her needs in having and raising a family. Although she was able to stay mathematically active as her children grew up, it is certainly plausible that she would have published more in her youth had she not had these additional responsibilities.

Ironically, one argument used to explain why age hurts mathematical productivity is that one becomes more distracted with age and loses the ability to be completely focused for hours at a time. But for many of these women, the opposite was true. Those with children found that their research focus was much improved as their children got older and demanded less of their time. They often relished the opportunity to be more immersed in work without having as many domestic responsibilities to juggle. In this way, some women with children become less distracted with age rather than more.

But other women, such as Lenore Blum, felt that having a child did not slow them down mathematically. If anything, it was the *assumption* among Blum's peers and teachers that it would slow her down that was the problem. And indeed, as we have seen, the assumption (rather than the reality) that mothering and mathematics are incompatible can be problematic for women. Is this assumption even accurate? Zuckerman and Cole have examined this question in science more generally. They found that contrary to popular belief, "Married women publish more than single women, and mothers publish more than childless women. Furthermore, women's rates of publication, on average, do not decline following childbirth and during the years when they are caring for young children."[14]

There may be no comparable study for women in mathematics, but it is clear that at the very least, we cannot automatically assume that marriage or motherhood is incompatible with mathematical productivity. The relationship between the two is complex. For example, timing is a critical factor—having children during the relatively flexible years of graduate school may be easier than having them during the highly pressured early years of a tenure-track job. The kinds of institutional and departmental support one receives can also make a tremendous difference in one's ability to balance work and family responsibilities.

Whether or not having children translates into lower productivity, the reverence for youth can have a detrimental effect on women in a variety of ways. For example, some women put off having children until later in their careers—until after they have tenure or are well established in the mathematical community. Some find this choice ideal; others are disturbed when they find it difficult or impossible to conceive children later in their lives. Still other women may choose not to go into mathematics, or to leave it prematurely because of the perceived difficulties in balancing having children with a mathematical career.

More information is clearly needed to understand how having children influences one's mathematical life. But equally important is what the mathematics community does with that information. If women's productivity *is* found to decline for a few years while their children are young, is this information used to discriminate against women, or is it used to determine ways the mathematical community can accommodate the having of children? As Karen Uhlenbeck says, "I think that the system is not set up to handle women taking time off to have a family. That's because it's the structure of the system. I don't think it's a structure of doing mathematics."

The ideology of youth can certainly be problematic for men as well as women. It leads to an atmosphere in which mathematicians often feel that they have little to contribute to mathematics once they have passed their supposed prime. This kind of pressure can simply contribute to insecurity rather than cultivating productive development. It inhibits an appreciation of the contributions that mathematicians can make at all stages of their lives.

Why Does the Image of Youth Persist?

Given that the facts do not overwhelmingly confirm the image of a mathematician as productive only at a young age, we are left wondering why this image persists. A partial answer may be found in the relationship between youth and some of the prominent images of mathematicians discussed in earlier chapters. For example, the image of youth may be tied to the image of a mathematician as a loner. Perhaps mathematicians who work in isolation *are* more likely to be increasingly less productive over time, while for mathematicians who work closely with others in the mathematical community, youth becomes a less reliable predictor of productivity. Thus, different components of the imagery and ideology of the mathematics community can have a self-reinforcing effect.

The image of youth is also tied to the vision of a mathematician as a young, virile male. The common reference to mathematical "vigor" or "virility" emphasizes that what is being measured is a kind of potency often associated with masculinity. We are led back to the image of the mathematical cowboy. As D. E. Smith wrote, "Geometry is a mountain. Vigor is need for its ascent."[15]

The centrality of youth is also tied to the image of a mathematician as a kind of mental athlete. If the prominent belief is that mathematical ability is largely innate, and that it peaks at an early age, then decline in later years is seen as inevitable. In this vision, mathematical talent is like a flower waiting to bloom, but once it has displayed its glory, there is nothing that can bring it back to life. Such a vision creates little incentive for finding ways to keep mathematicians productive throughout their professional life.

This persistent belief in youth despite the numerous examples to the contrary is fascinating. As Bela Bollobas reminds us:

> Littlewood remained active in mathematics even at an advanced age: his last paper was published in 1972, when he was 87. One of his most intricate papers, concerning Van der Pol's equations and its generalizations, was written when he was over seventy: 110 pages of hard analysis, based on his joint work with Mary Cartwright. He called the paper "The Monster" and he himself said of it: "It is very heavy going and I should never have read it had I not written it myself." His last hard paper, breaking new ground, was published in the first issue of the *Advances in Applied Probability*, when he was 84.[16]

Clearly mathematicians can, and do, make important contributions at all stages of their lives. Indeed, as the mathematician J. J. Sylvester points out, mathematicians tend to exhibit "extraordinary longevity": Leibniz, Newton, Euler, Lagrange, Laplace, Gauss, Plato, Archimedes, and Pythagoras all lived and were productive until at least their seventies or eighties.

Though this focus on youth is certainly true of other disciplines, it is more acute in the sciences, and particularly in mathematics. It is reinforced by the glorification of young prodigies, and is tied to the belief that mathematicians are born and not made.[17] Indeed, the few known prodigies are dazzling; their mathematical prowess before they have moved beyond puberty is astonishing. Sometimes that talent is specialized (e.g., in doing elaborate mental calculations or identifying huge prime numbers); other times it blossoms into a deeper and sustainable mathematical maturity. But few traits (except perhaps musical or artistic genius) exhibit themselves in such a startling way when one is still a child. Ironically, such prodigies were historically treated quite badly, often seen as freaks of nature or evil omens. But today they are most often glorified as mythic heroes, for they clearly play into the imagery and values held in highest esteem in the mathematics community.

The perception of prodigies is that their abilities shine extremely early and then burn out early as well. But in reality, only in a few cases does that ability erode quickly; and those cases usually involve isolated sorts of abilities that disappear before or during adolescence. Most who maintain their ability through adolescence continue to have it throughout their lives. Only in some cases does it translate into deep and mature mathematical talent, as in the case of Gauss. But Gauss continued to be mathematically productive throughout his life.

In summary, we see that the image and the reality of a mathematician may not be the same. In this case there is a deeply entrenched belief that mathematics is a "young man's game," despite the fact that there is no compelling evidence to support this hypothesis; indeed, the studies that have been done suggest the contrary. But we also see that when the image and reality differ, it is often the image that can have a more powerful influence on attitudes, practices, and policies.

These attitudes about mathematics and youth have significant practical ramifications. If the focus were not so much on the young, virile mathematician, it would be easier to design programs with women in mind. For example, it would involve accommodating the fact that women are likely to have children in what is traditionally considered their prime mathematical years, and looking at their productivity over a longer time span. It means recognizing that women may need to enter the mathematics research pipeline later in life, as Joan Birman did, or they may need to work part-time for a period to balance having children with mathematical research, as Mary Ellen Rudin did.

Accommodating women could mean cultivating a multiplicity of models for a mathematician's timeline, rather than seeing a mathematician as a kind of athlete who peaks in his or her twenties and then burns out over the following decades. It also involves creating many entry points into mathematics, and many avenues for reinvigorating mathematical curiosity and productivity.

Finally, less of an emphasis on youth could also lead to a somewhat different set of values. Celebration of mathematical maturity rather than just youth could lead to more recognition of the wide range of ways mathematicians contribute to mathematical knowledge as a whole. This includes, among other things, stimulating other mathematicians, suggesting good problems, making connections between fields, and providing a larger vision of the history and context of mathematics. A theme that was prevalent in these interviews is that the women saw age as an asset—it gave them a sense of a larger picture of the mathematical landscape, a better sense of what kinds of problems were important, a wider range of techniques to draw from, and a deeper appreciation of (and insight into) how different ideas fit together. As Karen Uhlenbeck puts it, "I still get a kick out of doing mathematics. It's harder to come by now because many of my standards are higher. But now there's a new kind of pleasure of trying to fit things together, making something match something else. When I was younger, I had no in-depth knowledge of mathematics. By now I know an awful lot of mathematics and I'm really fascinated by connections."

In the end, as these interviews and other studies have documented, what seems far more critical than age for mathematical productivity is whether one is in an intellectually stimulating environment.[18] It would seem, therefore, that the mathematical community would be well served by examining how to cultivate environments that feed enthusiasm and stimulate research.

Joan Birman with oldest son Kenneth,
on the beach in Provincetown, Massachusetts

Joan Birman at Barnard College, 1985

Joan Birman

(1927–)

Joan Birman grew up with her three sisters in Lawrence, Long Island. Three out of the four sisters majored in mathematics in college, though only Joan went on to become a research mathematician. After having three children of her own, Joan went back to do graduate work in mathematics at New York University, receiving her Ph.D. at the age of forty-one. It is remarkable in mathematics to get a degree so late and then to go on to become such a productive and successful mathematician. Joan Birman exemplifies one way to integrate family with a research career; her life gives us insight into some of the advantages and disadvantages of such a path. Like many of the women in this book, she challenges the myth that "mathematics is a young man's game."

Joan's research is in knot theory. She teaches at Barnard College–Columbia University and gives talks in mathematics all over the world.

From Tinkertoys to Topology

What did your parents do?

My father, George Lyttle, was a dress manufacturer. Neither of my parents finished high school. About the time I was married, which was 1950, my father retired because of economic problems in the dress industry in New York. He intended to take up some other area, because he was then only in his mid-fifties, but he never did. I think my father, under different circumstances, could have been a scholar of some kind.

Was either of your parents interested in mathematics?

No, they didn't finish high school. My father had to work for a living from his very earliest days, and he just didn't have the opportunity to think about other possibilities. Both my parents came from a generation where women did not think about such things. But I'm not sure my mother was inclined that way anyhow. But both my parents had a strong idea that we would all go on to college.

I am one of four girls. I am the third child. One sister, the one who is just older than me, and I had almost parallel careers. We both brought up our children and then began school afterwards. She studied plants and I studied mathematics. We both got our Ph.D.'s when our children were grown. I was forty-one; she was about the same age, a little bit older. She was four years older than me. We both had successful, recent academic careers in research. She was a plant physiologist. She died in 1989; her name was Ruth Lyttle Satter. Both she and my oldest sister were math majors in college. So three out of the four girls majored in math.

Did your father show interest when you were in school?

Always. Neither one of them ever had any interest in the specifics of mathematics. But if I came home with a 98 on schoolwork, my father would say, "What happened to the other two points?"—and at the same time indicate how pleased he was.

In some ways neither my mother nor my father understood that mathematics was not a woman's world, because they didn't understand enough of the subject. As long as you were studying, that was good, that's what was important. In fact, as time went on and my various nieces and nephews went to college, my parents never discriminated against the one who was a photographer or doctor or the one who was working on a publication on women's health—as long as you were at the books, it was good.

And doing well.

That was not an issue; we all *wanted* to do well. One way my parents encouraged me was with the toys they bought for me when I was very young. They bought me an enormous set of Tinkertoys with wooden sticks and connectors that you could use to build large structures, and also an erector set where you put pins into hinges. They got me other toys as well, and I would put things together. I'm a topologist, and all those toys involve shapes and structures just as topology does. I had a chemistry set too. Maybe my parents recognized my interest by choosing toys like that. They didn't push what should be a girl's choice on me. As long as I was really interested, it didn't seem to bother them that this was not the usual choice for girls.

Did you and your sisters influence each other in mathematics?

Hard to say. I like to think that I made my own choices in later life, but I have to say that the evidence is that we probably did influence each other, sure. I was the third.

Do you remember them talking to you about mathematics?

No. My early memories are that I was good at math and could understand more than other people. And I liked it. I've always wanted to understand things.

Mathematics and other subjects?

Mostly mathematics. I remember mathematics, specifically, when I was very young. I remember being able to figure out something about the sum of two numbers. It all came back when I went to visit my children's school and I heard the teacher giving a lesson about whether the sum of two odd numbers was odd or even and what happens with the products. I remembered it from when I was a child, and I remembered how I understood it right away and nobody else did. And I remembered how beautiful it seemed.

So elementary school was where it started for you?

Yes. But I don't remember having any particularly good teachers in elementary school. I do remember a good one in high school.

What do you remember about that high school teacher?

We had a group in high school—myself and three or four other girls who loved math—and this teacher was really positive. It was a girls' school. We would just sit there, and our arms were almost coming out of the sockets trying to answer the question. We didn't need encouragement. We were competitive with each other, and she was just a good teacher, and the material was interesting. Geometry was a course that I loved in high school. This group of friends and I went around the school to recruit candidates for a solid geometry course. We succeeded, so the school gave us a solid geometry class.

Julia Richmond was a great big high school, but it had a separate little school within the school, called the country school, which was for academically strong students. When I first got there, I was not in the country school; I was admitted to it afterwards. It was a very nice school. Within that group it was possible to love math and not feel like an oddball. It wasn't until later that I began to think there was something a little inconsistent between being a woman and being a mathematician.

How much later?

In college, when I became interested in boys, which I wasn't in high school. They weren't around in high school, there was no opportunity, and I thought that the girls who were interested in boys were silly. I guess I matured rather late.

So by the time you were in high school, you were really excited about math.

Yes. And I wasn't concerned about the fact that this set me off in some ways from most of the other students. There weren't that many girls who were into mathematics, but I had a good circle of friends, and that was enough.

At that point did you think that you might pursue mathematics later?

Sure I did. Then two things changed that. The first was that college mathematics was initially quite disappointing. I started with calculus, and I didn't like it or have enough sense to understand that it was the course and not me. It just got to a point where I felt like they could tell me anything and I'd have to believe it. All my confidence in my knowledge and understanding of math was gone. Most of the students that I teach now are happy with that kind of a calculus course; in fact, we have a hard time getting students who want anything different from that. But I found it very dissatisfying, and I didn't have any idea what it was that I didn't like. It seemed like mathematics had changed. That was the first thing that led me to question a career in mathematics. But later, when I understood things better, a second issue arose. I became aware of the fact that mathematics required enormous concentration, and that if I was going to do it badly, then I might as well not do it at all. That's when I decided not to go on to graduate school. I knew that it required a kind of concentration that I felt was going to interfere with the rest of my life [as she describes later, her focus on relationships: marriage, children, etc.].

Were there other subjects that you were particularly interested in?

I was a good student. But I knew I had no talent for languages whatsoever, and I had a hard time remembering history. With mathematics, once I understood it, I didn't have to memorize it. My interests were certainly in the direction of science and math. I liked biology. When I went to college, I somehow thought about physics, biology, astronomy—astronomy was very, very interesting to me. But in picking a career, I thought that with astronomy you have to live in a place where the sky is clear enough to look at it—that didn't sound consistent with the city life I liked. I liked other things: sewing, cooking, things with my hands. I was clumsy and I knew that I wasn't going to be any kind of an athlete, but I like doing things with my hands. I was never somebody who could take being with people all day long.

So I majored in math in college [at Barnard]. I didn't go to graduate school right afterwards. I worked. In fact, I got an initial job in a place that was making electronic equipment. That job lasted about six months. Initially it was very interesting; it involved solving a problem in geometry. But after the problem was solved, they had me making measurements on the oscilloscope, and that was terrible. I figured that there weren't too many jobs like the first one I had, so I'd better learn something more practical. That's when I started to go to graduate school [in physics].

I took a lab course in electronics at Columbia in the Physics Department, which I liked very much. I like working with my hands. Then one day—and this was really an accident—I met my old physics professor from Barnard, and he asked me what I was doing. He said they had an emergency and needed a teaching assistant for the physics lab at Barnard. He suggested I go back to graduate school and take the teaching assistant job. So that's what I did. That's how I got to graduate school.

What a coincidence.

Yes, it really was. I did finally get my master's degree in physics. But by then I knew that I didn't have the talent for physics. I scraped through, but I didn't have the feeling of it. Again, I felt like they could tell me anything. I didn't understand what the ground rules were.

And intuition.

Yes, you need some kind of an intuition, and I didn't have it. The electronics laboratory involved a very precise measure of truth. I liked that. But when I got to problems in mechanics, I just didn't understand what you could ignore and what you had to accept as given. There always seemed to be approximations. But I never knew which ones were acceptable, and the whole thing was hazy.

By then I had a bachelor's in mathematics and a master's in physics and this one year of job experience, and so I went out to look for another job. I got a second job in the aircraft industry working on early navigation computers for aircraft. That was very interesting. I did that for five or six years. In the meantime I had gotten married. I held that job until we had our first child. I had intended to go back to the same kind of work after we had children, but I found that it was just impractical. I didn't want to. I did work part-time, which was important to me, but it got to be more and more difficult to make it meaningful. So after some number of years of working one, two days a week at the most, there was a crisis. My husband, who worked in industry but had leanings toward academia, had an offer to be a visiting professor at the University of Pennsylvania. It meant moving the whole family and leaving my part-time job. I was away from the job for six months, and when we returned, it just seemed impossible to go back to the same kind of part-time work.

While we were at Penn I took a course in digital computers, and that was interesting. While I was working, I had never felt like I knew enough mathematics; I wanted to learn more so that when I returned to work I would be better equipped. So after our third child was born, I started graduate school in mathematics. Throughout this period, then, I never really lost track of mathematics; I always kept some little thread of contact, even through the time the kids were growing up.

Graduate School

I took one graduate course in the evening at NYU, and that went so well that the next year I took two. There was an exam that determined who got a master's and who was able to go on for a Ph.D. So I took the courses that were needed for the exam. By then my three children were a few years older. We had a babysitter take care of them during the summer so that I could study. I spent the whole summer in my bedroom office studying for the exam, thinking it would be for a master's. But to my surprise, I passed it for a Ph.D. qualifier, and I was very pleased!

So it was the same exam? It just depended on how you did on it?

That's right. I was surprised and pleased. At that point I thought if I really wanted to go on to get my Ph.D., I needed more help at home, and I couldn't afford it, so I applied for a fellowship. By then our youngest child was in nursery school, and I was awarded the fellowship upon the condition that I would come back as a full-time student. So that's what I did. I used the money that I got from the fellowship to get somebody to come help. At first it was very good. We had a lovely woman. She really added a lot to the household.

Did she live with you?

No, she didn't live with us. We never had anybody live with us. You know, studying mathematics, the amount of time I had to be in school was really relatively little, but what I needed was time to be free to work, so we used to have people who took care of the children and who cooked, varying hours, varying arrangements. I was forty-one when I got my Ph.D. As soon as I became a full-time student, I started to work with other students, and that made a big difference.

Did you work with women students?

No, the ones I worked closely with were all men.

How many years were you actually at NYU?

Our son, David, was born in January of '61, and I got my degree in '68. So it was seven years altogether. At first the pace was slow, and then it picked up.

That's great that they gave you that support, because otherwise you wouldn't have . . .

Couldn't have made it—it really was important. I feel very grateful to NYU for that. First of all, they had this very open program for people working toward a Ph.D. You could come as a part-time student. That is not true at all graduate programs.

Who was your thesis advisor?

Wilhelm Magnus.

At what point, then, would you say you really got turned on to the research mathematics? Was it in graduate school?

No. I had worked on research problems in industry and I liked them. Those problems had the same quality as the mathematics that I'm doing now. But when I started to do research for my Ph.D. it grabbed me, because it seemed as if the purpose of the whole thing was much more beautiful to me than designing a new piece of aircraft equipment.

In industry, I had worked on very applied problems. For example, I worked on a Doppler navigation system in one of my jobs. I worked on another problem—this was on a bombing computer, and I really did not like the purpose of it, but it was a problem where I was very proud of my idea. The problem was to compute the effect that the wind and an aircraft's up-and-down movement would have on a bomb when it was dropped from the aircraft. The trajectory of the bomb could be described with differential equations. They asked me to figure out how to correct the equation to account for the motion of the aircraft, but what I realized was that all we had to do was change the initial conditions—the differential equation would stay the same. So even though the change was a change in wind speed, you could simulate it by a change in the initial position. I had a hard time getting them to understand that. I had such an argument with my supervisor because he could not follow what I was saying. And I was right!

That's an interesting difference between research work and working in industry. In industry you've got to convince somebody of your explanations, too.

Yes. So when I started to do research in mathematics, it seemed as if it was more meaningful to me to make contributions to mathematics than to navigation computers. People ask what the use of mathematics is. I had this whole long period when I was doing very useful things in the ordinary sense, but then I found it more meaningful to be contributing to [pure] mathematics. Most people feel the opposite.

First Teaching Job

I knew I wanted to do research. The job market was very poor when I got my Ph.D. Also, I was a woman, and there were very few women in mathematics; I was older, and there was a lot of prejudice against that. It looked like I wasn't going to get a job. I was restricted geographically; because we had children, it was impossible to go anyplace far from home. So I was limited to all the colleges in the New York area. I was offered a job in one of the branches of the city university with a high teaching load and uninteresting courses. I didn't want to do that. I really wanted to be a research mathematician.

Then I thought, Why couldn't I just do my research and attend seminars and keep in touch that way? I went to a meeting, and just by accident people had heard that I was looking for a job. Stevens Institute, which is an engineering school in New Jersey, just happened to need people. It was late in the year, after the academic hiring had been done; several people had left suddenly, so they had vacancies. So by almost an accident, I got that job. And it was a good job. It was an engineering school with good students; there was always somebody who responded and kept you on your toes if you did something stupid in class. . . .

I was there for five years, with two interruptions—one was when my husband had a sabbatical leave and we all spent a year in Paris; the other was the year I was teaching at Princeton. That came about through my research. I had done some good work, and I was invited to give a seminar at Princeton. The following week I went to attend the seminar again, and Ralph Fox, who was a very well known mathematician in knot theory, which was what I was interested in most of that time, said, "How would you like to come and visit for a year?" I talked this over with my husband and he said, "If you want to do that, we can move to New Jersey and I'll commute." But the three children had just had a year of their schooling interrupted when we were in Paris, and I knew we'd have to make arrangements to rent a house, find a place where my daughter could take swimming lessons and music lessons, deal with transportation, etc. I couldn't handle that, and the children were sensitive to these moves. So I thought I would commute.

I knew this was my chance to be a research mathematician, and it was. It made a tremendous difference—because of the prestige of being at Princeton for a year, the context, and everything else.

You have had many fortunate twists in your life.

Well, yes, I think that's true. A lot of fortunate twists, but I think you have to be alert to them.

Yes, if it weren't for those, it might have been something else.

I think so. I do a lot of joint work, and people ask me how I find somebody to do joint work with me, and it always seems like it's almost accident, but of course it isn't. I am alert and ready for it. I'm looking for it.

Choosing a Field of Research

When you started graduate school, did you know what field of mathematics you wanted to go into?

When I went back to graduate school, I was fully intending to get back into the aircraft industry or some math-related job. I picked NYU because it was a school of applied mathematics. It was completely surprising to me that I became so interested in pure mathematics. That happened by a process of discussion and learning about things. After I passed my Ph.D. qualifier exams, I went to speak to different people on the faculty about what they were doing, and Louis Nirenberg said something that was excellent advice to me. I had liked him. He was an excellent teacher. In fact, he taught a course that was really important to me. I said I didn't know what area I wanted to work in. He said, "Do you like inequalities?" And I said no. And he said, "Well, you don't want to work in differential equations."

That was very good advice. At NYU there weren't too many people who were not in applied mathematics and in something related to ordinary or partial differential equations. One of them was Magnus, who was ultimately my thesis advisor. He worked in combinatorial group theory.

Then how did you get into topology? When did that transition happen?

The two subjects are really very closely related. By the time I'd gotten my Ph.D., I knew that low-dimensional topology was the thing that was more interesting to me than combinatorial group theory, and I just gradually worked my way into it. I made some contributions and used a little bit of topology. In my thesis I solved one problem, but it suggested another problem. This is what I was really ambitious to solve when I was at Stevens. In one of the nearby offices there was a young fellow who had just gotten his Ph.D. at Stevens, and he, like me, was a little bit beyond the usual age, though much younger than me. He had been an engineer and he didn't like what he was doing, so he went back to graduate school, and he had just gotten his degree. Because he got his degree rather late in the year, they kept him on as an assistant professor. So we began to have lunch together. I talked to him about this problem that I had tried to solve. We talked about it through the whole year. I had a conjecture, but first the conjecture didn't make any sense, and then gradually we began to understand some of the structure that I was guessing was there, and we began to understand it better and better. Then came a key moment, and I remember it very well—when he put something down on the blackboard and we started going on it. We solved the problem together, and that's what I was invited to give a seminar on at Princeton. It was a very good, satisfying piece of work.

Was that your first joint work, also?

That was my first joint work.

That's a nice way to do it—just sitting having lunch and working on it a little bit at a time.

Yes, that's right. But all my joint work has been like that. I just talk to people and it just happens.

So that's when you really started getting more and more into topology?

Yes. And then I gave a talk on our work at Princeton, and I discovered that there was this weekly seminar on knot theory. My work involved surface mappings, but surface mappings are very closely related to knot theory, so I began to attend this seminar on knot theory at Princeton and began to go to it regularly, and that's how I learned topology.

Did you stay in contact with people at Princeton and continue to work with them?

The year that I was there, sure. After the first year, Ralph Fox started to have health problems and ultimately had a second heart surgery. He died a week after the second surgery. After that I felt so bad. The whole seminar fell apart after his death. But by then I had a job at Columbia, so my contacts at Princeton ended because the graduate students who were working with him were no longer there. There wasn't anybody else who was doing just what he was doing.

That must have been rough for the graduate students, too.

When he died, he had one graduate student who was still working with him. When he knew he was going to go in for open-heart surgery, he asked me, "Will you look after him in case I have any problems?" In fact, this graduate student applied for a job at Columbia, and I really wanted him to have him there. I was very touched that the department respected my wishes and offered him a job.

So you got a job at Columbia right after Princeton?

Yes. I was doing good work, and the year at Princeton was very helpful to me. People knew about what I was doing. The very process of my possibly getting a job at Princeton made my name known, so it was quite important. I was offered a position at Barnard-Columbia [two schools with one mathematics department].

A Late Start

When you were going to graduate school, you had an unusual background since you had worked for a while first. Can you talk about the ways in which that made your experience in graduate school different? Did you feel more mature than other students in your attitude about graduate work?

More mature in a sense that I knew what I wanted to do. I was glad to be in graduate school, and I wasn't fussing like all of the other students were. I was past the adolescent angst. In college I was very concerned about other aspects of my life, of my relations with people and where I was ultimately going to live, would I get married, what was I going to do with my life. When I was a graduate student, my personal life was somewhat settled. I had a good, secure marriage, and I had had a good shot at being a housewife. I couldn't see that as a lifetime occupation. It didn't interest me.

Mathematics did.

Mathematics did, and that was enough. The other students, on the whole, the younger ones, were much more preoccupied with their personal lives than I was. They had the luxury for that; I did not.

They had the luxury for it, but on the other hand you had the luxury of not having that distraction.

I didn't have that distraction, but I had plenty of other distractions with the children. I had a lot of responsibility. The other women I knew could not understand how I was able to do it. But the moment everybody left in the morning, I sat down at that desk, and nothing interfered with my concentration. I really worked. The way I did it is that I was interested. It wasn't hard. I was enjoying it.

But it was hard initially to get back into it?

I had always had a little thread of contact. There wasn't a discontinuity like that. And we didn't really need the money; our needs were simple, and my husband's salary sufficed. When I began to work, I was damn glad to have it, and it added something quite extra that I didn't think about, a certain independence that was very nice. But I liked what I was doing so I had no conflict.

Were there ever times in the beginning where you were discouraged and felt like you weren't going to go on?

No. I didn't have any real crises like that. I knew that the other students did, but I didn't.

I'm wondering if that was because you entered research mathematics relatively late.

I think that was a good way. It filled an enormous need in my life at that time that I got started in research mathematics, and it's continued to. Especially because my husband has had a very active career, so he's always busy with his own thing, and I've got to have my own thing. Otherwise I would feel like an appendage to him. My work is necessary for me in order to be myself.

Did you have mentors in either college or graduate school that were particularly important to you?

People were certainly helpful. Magnus was, and Fox was. But I had enough confidence on my own. But I don't think that made an awfully big difference. I tend to discount this role model idea. I think that if a person doesn't have enough underlying feeling of "I'm going to do what I like" and not be so affected by what other people think, then they're not going to be mathematicians anyhow. I certainly knew that being a mathematician was not what the average man on the street could comprehend. But I felt right at home with mathematicians.

Relationships, Marriage, Being a Woman

You haven't talked very much about college—high school and graduate school seem more significant in your mathematical development.

That's true. In college my preoccupation was with other things, i.e., my relationships with people. That really took precedence over the mathematics.

Is that when you met your husband?

I met my husband after I finished college, but there were other people that I was dating in college, and that whole issue was very much on my mind. Not only did I know that I had to or wanted to get married, but there was even a feeling [in my family] that my sisters and I should all get married in the right order.

Did that happen?

Yes, it did.

I could see why in college relationships would be such a big thing. For one thing you were at an all-girls' high school, so that was the first time you met men.

That's the first time I began to date.

Where did you go to college?

I started college at Swarthmore but didn't like living in a college dorm. So I transferred to Barnard and lived at home. I was very happy to be able to do that. My sisters had done the same, so that model was there. I felt better with the privacy of living at home. I really have a limited ability to be with people.

Did you get married soon after you met your husband?

A year and a half later. I graduated from college in 1948. I was married in 1950. It was a stormy period. It's a very difficult decision to make—is this the right person for me? It just took all my attention. I guess that one of my thoughts about women in mathematics is that it is just a much more absorbing issue for women at a time when men say this is when you do your work or you'll never do it. And we women buy that hook, line, and sinker.

I see it in the graduate students. There's just another pull that the women have; it has to do with home, family, human relationships, etc. And while the men are also involved in all those things, it doesn't seem to take their attention quite the way that it does for women. I don't know. You're from another generation, and I always suspect that this is just my experience.

The bringing up of a child takes an enormous amount of attention, and you don't want to put it aside. It goes so quickly, and if you're not there, you miss it. It was different for my husband. He was working so hard, and had pressing responsibilities to "take care of us," all four of us. Even when I started to work, the personal issues were still there. However, I did bypass many of the problems by starting graduate school later, when the children were older and the demands were much less.

Yes, this is a primary conflict for women, particularly in academia, because often women are having children around the same time they're coming up for tenure.

Yes, and the men are working like lunatics on their mathematics, and the women, they see this choice, and if you just give it a little less effort, your research is dead. I think that's a big issue. I guess that if I could see any solution to the non-participation of women in mathematics, it would be, first of all, if women were able to think about going back to mathematics at a later point, and if there was a practical way for them to do this, and if you could reach the right women, and if the whole community was ready to accept this as another option—all of these things would help. But the fact is that the community has bought hook, line, and sinker this whole idea that if you are going to do research, you do it when you are young.

Do you think that there's any merit to that? Why is it such a powerful myth?

I don't think there is merit in it. I think doing math when you're enthusiastic, yes, that's what is important. Not your age.

Women, Mathematics, and Children

When you look back, in what ways has being a woman affected your career?

It affected it enormously. I took a fifteen-year break. I got my Ph.D. when I was forty-one, not twenty-five or twenty-six. That's a great big difference. It's affected my life in absolutely fundamental ways. It's made it hard to be a mathematician in some ways. It was lonely out there in 1968. I was not crazy about that. It would be a different world if there were lots of women mathematicians. We would feel very differently. But I don't know; maybe I like aspects of that.

Do you think it is easiest to do the children first and then go back, or are there ways women who are having children can be accommodated in the system?

I don't know the answer to that. I think it's something that each person is going to have to suffer through. It may be much easier now than it was in my day, because I think that a lot of the younger men are much more understanding of this problem and that the upbringing of a child is not a woman's responsibility as it was when I was brought up. It's so common now for women to be working.

So did you use nursery school?

Yes, we used nursery school, but in those days nursery schools were not designed to help the mother. It was impossible for me to have any kind of a career without having anybody really in charge at home.

How old was the youngest when you started?

When he was an infant, I started going to school at night. My husband was the babysitter. I found that easy. I found that I had enough time of my own so that without asking anybody to help out, I could handle three children and study for that one course. But then the second year, when I took two courses, that was more ambitious, and that was really hard without help.

By the time you were a full-time student, your kids were in school, so there were regular hours. Did that make things easier?

When I started being full-time at NYU, my youngest child, David, was starting kindergarten. But even kindergarten was challenging because it was only a half-day. And a full day of school was until 3:00 P.M., and no afterschool programs existed. Quite the contrary; it was a suburban neighborhood. There was a lot of carpooling and taking them around to visit one another, and I was very hard-boiled about that; I told my children that they had bicycles and feet!

So were you really unusual in this neighborhood, being a woman working outside the home?

Yes, it was isolating, absolutely! My involvement with the community went way down when I started to be a serious professional. Still, I think that what I did was easy—the advantage I had was that I did not have a Ph.D. before we had children. So I got my training and I used it right away. I think the people who have the biggest problem of all are the ones who get their training and then feel that they've forgotten it all and lose their confidence when they are ready to go back to work.

The way that I got myself back in was quite natural and gradual. I did it a little bit when I had a little bit of time, and I was slowly able to build it up. But somebody who was working at a peak and then found at a later point that bringing up a child is really time-consuming and takes a lot of attention may feel like they are losing touch with the field, or feel like the job is being juggled all the time with child care. That kind of a person is going to have a very different feeling when the children grow up, whereas I was starting something new. I wasn't going back to something old and feeling like I'd missed many years.

Right. And the enthusiasm you had because you were starting something new really carried you, too.

That's right. And I also was able to build it up very slowly in a way that was quite easy. If it's mathematics that you're doing, a lot of your time is spent with books and paper. You don't need a lot of equipment. You don't have to be long hours in a laboratory. So I could be at home when the kids came home from school, and they would just want to come into the house and say hi and run out again. I could spend a lot of time just doing that, and it didn't really interfere with my ability to study. So in some ways that made it very easy. I don't know if that's a model that would work for anybody else. I also wonder how to reach the people for whom this would be of interest. How do you find the women whose children are growing up and who feel they could get back to mathematics?

I think that NYU was really excellent in this respect. The graduate school was open to the community, and it's an excellent school. The atmosphere is really stimulating and very good, and there was a constant flow of people from all sorts of different circumstances who were doing serious graduate work, and you were also with students who were full-time students. In 1996, Columbia still has nothing like this for graduate students.

A graduate school that has a good master's program at least has some way for people to do a little bit. You see, I never planned what I was going to do. I didn't really set myself a goal, like wanting to be a research professor in mathematics. I had very small goals. I wanted to be able to learn the material of this linear algebra this semester, and then it built up after a while. That was very handy.

That seems very common among women, that they have small goals first and then they keep going.

You have to feel your way.

And it's not a given to many women that they're going to have a profession. It's a given for most men. Women often do it because they love it, but they don't necessarily think (at least at first) of math as their profession.

Yes.

That's starting to change a little bit, but it is still predominantly the case. What you say about NYU and their willingness to be a little bit more flexible is very important.

Yes.

Encouraging Women in Mathematics

Over the years, you must have thought about how to get more women involved in mathematics. Do you have ideas about how to encourage them to do math, and then to stay with it?

They have to feel passionate about doing mathematics. Doing creative work takes a passion. You have to be driven. So what can people do to help? I really think that in some ways I've been helpful to others, but it's never been through committees and policies. You can do something in other ways too. Last year, one day there was a knock on my door, and totally unexpected a young woman came in. She was a graduate student at another university and was depressed and unhappy about graduate school. All of a sudden all of the starch had seemed to go out of her life. She used to love mathematics, and she was in the middle of writing her thesis. Why did she come to see me? I don't know. Somebody said to her, "Go talk to Joan Birman; maybe she can help you." I felt that I could remember myself at that time; it wouldn't have been a great time to be writing a Ph.D. thesis. I don't know what else was on her mind, but she wasn't getting the pleasure out of math that she had in the past. She was thinking of dropping out of graduate school and ultimately did not, and I think maybe I helped her in a way that a man would not have been able to at that moment. I think the fact that we were two women did make that a little easier to her.

Do you remember the kinds of things you said to her to help her out at that stage?

Maybe just that she was able to cry in my office and wouldn't have been able to cry in front of a man.

So it's really through individual cases that you feel you can encourage women in mathematics?

Individual cases, being alert, being there. I think it's a small thing.

The Nature of Mathematics

What do you think mathematics is all about? Are we discovering things, creating them? How do you think about what you are doing when you are doing mathematics?

My husband and I have had a big discussion about this, whether mathematics is real, and I guess I really do believe it is, very, very strongly—that there is structure out there and we're finding it. It seems to be endless. It's amazing. It's just amazing that there is always a deeper level that you can ask questions, and there is always a deeper structure.

Maybe mathematics is almost the easiest of the sciences. When I was in Monterey, California, I visited a marvelous aquarium, filled with fish and creatures of the sea who have the most extraordinary patterns and colors on their bodies. When you look at this, you think that nature is so complicated, and that mathematics is picking out the very simplest of the patterns and analyzing it. And that we have the easiest science.

Why are other sciences harder?

They're harder because the phenomena are so much more complicated that they don't admit the kind of sharp analysis that mathematical problems do. We reject anything that is too hard to understand. I just had a conversation with my gynecologist the other day, and it concerned hormone therapy. She described the different programs that have been prescribed for balancing how much estrogen and progesterone should be taken, and at what point in the cycle. She said you wouldn't believe the number of different studies of which way to do it. Nobody understands it. They don't understand fundamentally what is going on. Sometimes it's dangerous. People get cancer from it. They do not understand what they're doing. What do we do with a mathematics problem? We take something that's very simple and we analyze it. We pick problems that lend themselves to analysis. Doctors don't have that kind of a choice. Their subject matter is handed to them, and it's by nature very complicated.

Right. And in mathematics you can abstract out any little piece of it that you want. You don't have to look at the whole thing.

That's right. I think that in other sciences they try to do that, too. My sister Ruth, the plant physiologist, studied motions of plants. She had a plant that responded to a twenty-four-hour cycle. It folds up at night, and then the leaves fold in on the stem and the stem folds on itself. What she studied was not this phenomenon, because everybody knows it, but how the motion comes about. She looked at a particular membrane—potassium is transported across the membrane. Well, what is the mechanism that gets the potassium moving from here to there, and why does it happen? What's the chemical that gets it going? So it's almost like a mathematics problem. She tried to isolate, but no matter how much anyone tries to isolate, they may not be isolating the right thing. They don't know what the question is. But we, as mathematicians, can define our problems. We make our problems precise.

Jumping on the Bandwagon

That leads to the whole question of how we decide what are important questions in mathematics.

There are a lot of fashions in mathematics. Fashions come and go, and it's always a question of what do you do. Do you follow the fashion or not? Sometimes I tend not to because I dislike working under intensely competitive conditions.

Yes. It seems like you've been in it and out of it both.

Well, new things have happened, and when something big and new happens, then immediately there's a big rush toward it. I'm a little bit intimidated by that. I was in the middle of two big developments like that, and they both posed big problems in my mathematical life because the question was whether to try to get in there with the crowd; they were crises for me. Twice this happened where there was this enormous new explosion from somebody's work. Once was when Bill Thurston made the discovery that earned him the Field's Medal, and the second time was when Vaughan Jones did the same. The question for me was, Do I drop everything and rush in this direction, and if I do, will I contribute anything? Are other people better than me? Will they get there first? How do I handle it, and what should I do about my own research? Somehow after a period of anxiety, I kind of worked it out and came through the crisis.

When Thurston came along, there was a part of geometric topology which most people dropped. He brought a really new point of view, and the new point of view was geometry versus topology. Along with the new point of view came many new ideas that you had to learn, and the question was, Do I drop all the things that I know about very well and go follow the fashion or do I keep doing my stuff and pretend it didn't happen? Or am I just going to be left out? Eventually I kind of got over that and learned a little bit of the new and found a way to contribute.

When Jones's work came along, perhaps I made a bad decision because I was one of the first people who knew about it. The main discovery was made in my office, and I even played a role in it. At that time I was working with Carolyn Series. We were finishing up a paper, and the question was, Do I just drop Carolyn and drop this paper and run in a direction where I know I'm a little ahead of the crowd and have some advance warning?

What did you do?

I stuck with Carolyn and finished our paper, and I missed out on a whole lot of mathematics where I really had an inside track. I don't know if that was the right decision.

* * *

Birman's descriptions give us a sense of what life on the cutting edge can be like—the excitement and the intense competition, the tough decisions, and the unforgiving pace of discovery. In the end, she did ultimately return to work on the Jones polynomial and was able to "find her own niche" in the wave of activity that ensued. Indeed, Birman was able to find her own niche by paving a new path for integrating one's roles as a mathematician and as a mother—both of which were central to her identity. Though clearly this integration posed challenges at times, it illustrates that the two roles are certainly compatible once we are able to let go of traditional assumptions about when and how a mathematician's life unfolds.

4

Women and Gender Politics

Myth: Mathematics and politics don't mix.

The topic of women in mathematics is a controversial one in many mathematical settings, not only because it is about women, but because it is political. And for many mathematicians, mathematics and politics are spheres that do not, or at least should not, intersect. In this idealistic vision, mathematics is supposed to take us to "the farthest reaches of abstraction, where the cares of the world cannot intrude."[1]

But for many women, including several interviewed for this book, mathematics and politics are intimately intertwined. Indeed, for some, simply being a woman in mathematics is what *led* them to political activism. They did not start out that way—only when their expectation of equality did not mesh with the reality they faced did they become politically involved. It was not necessarily that their mathematical work itself became political, or even politically motivated, but rather who they were as mathematicians was inextricably linked to politics.

For the most part, those who did become politically active focused primarily on fighting for equal opportunity and equal treatment of women. Their activism usually took one of two forms. The first involved encouraging young women to pursue mathematics so that they would not be trapped by the stereotype that women can't do mathematics. These women activists recognized that most of the high-paying, high-prestige professions in science, medicine, and technology require a mathematics background, and by not studying mathematics, women were closing themselves off from major avenues of employment. The second form that activism took involved organizing women already in mathematics. The

Association for Women in Mathematics, formed in 1971, was a prime example of this kind of effort. It not only provided a sense of community and support for women in mathematics, who would otherwise be quite isolated, it also began to document inequity and advocate for change. Those involved in the AWM and other political activities were buoyed by the momentum of the women's movement, and highly motivated to make legal as well as social changes.

But women who became politically involved in the area of women in mathematics often met with a chilly reception not only from men, but often from women mathematicians as well. Some women mathematicians felt that it was inappropriate and unproductive to be politically involved in women's issues. As Mary Ellen Rudin adamantly said to Judy Roitman one day, "If you want to help women in mathematics, you do mathematics." She felt that focusing on the problems often makes things worse, rather than better.[2]

But Roitman disagreed. Though she did acknowledge that such work was very time-consuming, she did not agree that it was a waste of time. For Roitman it was a source of tremendous support, without which she might not have continued in mathematics or been able to develop a sense of community that helped her constructively address the kinds of difficulties both she and other women faced.

Lenore Blum and Judy Roitman are two women who were actively involved in gender politics. Though their mathematical research was in different areas, they knew each other well and played an important role in encouraging each other's mathematical career. They also collaborated extensively in their work for the Association of Women in Mathematics. A look at their lives yields a deeper understanding of how women come to be politically active, what role such activism plays in their lives, and how it integrates with their mathematical research.

Most of the women who have been politically active in women's issues have focused on policies and practices—and to some extent attitudes—of the mathematics community or society at large. But few have gone on to examine assumptions and beliefs about mathematics itself. In part, this is because most women mathematicians do not see traditional visions of mathematics as problematic; indeed, they tend to embrace them. But it is also true that to challenge the dominant vision of mathematics itself means risking being ostracized from the mathematics community. Simply being a woman defines one as an outsider to some degree; challenging basic assumptions about the nature of the discipline could constitute professional suicide. Only recently have some scholars begun to examine the fundamental assumptions about the nature of mathematics, and the impact of these assumptions on women's lives. In chapter 6 these questions are explored in more depth.[3]

Lenore Blum with husband Manuel and son Avrim

Lenore Blum starting position at MSRI,
1992 (photo by Arlene Baxter)

Lenore Blum

(1943–)

As Margaret Rossiter describes in her book *Women Scientists in America,* after the initial push to create and bolster women's colleges, which occurred up through the early decades of this century, there was a renewed backlash against women's presence in science. Two strong forms of discrimination emerged that served to "keep women in their place": either women were channeled into stereotypically female subspecialties of science (for example, home economics or botany), or they were hierarchically contained in the lowest tiers of a traditionally male field such as physics. Those few women who did manage to break through these barriers were seen and treated as exceptions. These women typically did not adopt feminist strategies or perspectives. Most often they chose to ignore gender bias and were grateful for the acceptance they did receive. This was a very pragmatic strategy, for it would have been professionally devastating to organize as women in science. Indeed, they were generally quite isolated as women in science.

A similar pattern emerges for women mathematicians during the middle decades of this century. They were, for the most part, isolated from other women, and were not inclined to focus on women's issues. Making an issue of being a woman in the mathematical world meant not being seen as a serious mathematician.

Lenore Blum, however, represents a break from this tradition of isolation. She is symbolic of a generation of women who came of age in the early days of the women's movement. Like many women mathematicians before her, Lenore began her career with the fundamental belief that she would be treated like her male colleagues—if she did good work, appropriate rewards would follow. Only when this assumption was shattered by the recognition that equality was more of an ideal than a reality was she driven toward political activism. Her personal experiences fueled her desire to create social and institutional change so that future women would be closer to real equality. In this way her life embodies the prevalent slogan of the women's movement, "The personal as political."

During the 1970s, Blum became a national leader in the field of women in mathematics. She helped found the Association for Women in Mathematics and became one of its early presidents. She was a dynamic spokesperson around the country on the topic of women in mathematics (a topic that almost no one was publicly addressing at the time), and created many new programs and institutions to promote and encourage women in mathematics.

But despite the tremendous amount of energy that Lenore poured into this cause, it is not what she had planned or hoped to do with her life. She always thought of herself primarily as a research mathematician. And throughout this period she hungered for the opportunity to return to her primary love of research in mathematics. It took more than fifteen years for her to be able to do that. Remarkably, at the age of forty, she returned to full-time research. She is now at the International Computer Science Institute in Berkeley doing research that bridges mathematics and theoretical computer science.

Why did it take her so long to return to full-time research? What obstacles did she encounter that led her to the "detour" of pouring herself into projects involving women in mathematics? And how was she finally able to return to research? In mathematics, such a return is almost unheard of; what strategies did Lenore use to accomplish this? How was this change received by the people around her?

Lenore's life is an interesting example of the tensions and balances between being independent and working in the context of community. Her story reminds us of both the advantages and disadvantages of being different, even marginalized, and at the same time it illustrates the importance of a supportive network that can sustain one emotionally, intellectually, and professionally.

The most important tool Lenore used throughout her professional life was forming communities. It was the mechanism that enabled her to give voice to the problems that existed for women in math, the mechanism that helped effect change both for women already in mathematics and in opening the doors of mathematics to more women, and it was ultimately the mechanism that enabled her to return to research in mathematics. In each case, creating a community was critical—it gave her a larger vision of her work and taught her about the power of collective action.

Early Period

Lenore's early childhood placed her in an environment in which being different was the norm. In 1952, when she was nine years old, her family moved from New York City to Caracas, Venezuela. Lenore, accustomed to the freedom of progressive New York schools, rebelled against the rigid classroom where she now found herself. Her mother, a science teacher, decided to keep both Lenore and her sister out of school their first year there. She educated the girls by traveling around the country, exposing them to a different culture, different customs, and different people. Adding even more excitement to the adventure, Venezuela was a political hotbed at the time, and the family had arrived during an attempted revolution. There were many American families who had moved there because of jobs in the oil industry, but it was a stark change from the almost exclusively Jewish community that Lenore had lived in in New York.

One important consequence of this move was that she met Manuel, the son of another Jewish family living in Caracas. Lenore was nine and Manuel was fourteen, and it was love at first sight. They saw each other off and on, as friends, for several years. Then, when Manuel went to MIT, they began a long and intimate correspondence that was to serve as a kind of courtship. Manuel's descriptions of long philosophical discussions and a community of intellectuals at the institute planted the seeds of Lenore's dream to join him and attend MIT. Her hopes, however, were soon dashed. Although she was an outstanding student, MIT denied her admission, claiming that it had enough dormitory space to admit only a tiny number of women.

What could have been an extremely discouraging blow became instead simply the first of several important detours in her life. Lenore decided to pursue her interests in both mathematics and art by studying architecture at Carnegie Mellon University (then known as Carnegie Institute of Technology) in Pittsburgh. Initially, architecture seemed like the natural marriage of these two interests; during her second year of college, however, Lenore realized that it was satisfying her love neither of mathematics nor of art. During this period it became clear that mathematics was where her real passion lay. So despite what her high school mathematics teacher had said—that she shouldn't pursue math became "mathematics is a dead subject, and everything was already proved more than two thousand years ago"—she switched her major to mathematics.

At the end of the school year, she and Manuel decided to get married.

It was 1961, and Lenore was eighteen years old. For some women, marriage signaled the end of school, but nothing could have been further from Lenore's mind. She transferred to Simmons College in Boston and lived out her dream of taking mathematics courses at MIT. It was the first time she found work that was truly engaging, challenging, and beautiful. During this period, her next major life dream took shape—she wanted to be a research mathematician. She was now more determined than ever to go to graduate school.[4]

MIT

MIT seemed like the natural choice for graduate school: Lenore was already taking courses there and doing well, and Manuel was already enrolled in the graduate program. Lenore went to talk with the head of the department, but the chair discouraged her from applying. Being a woman and being a graduate of a women's college were two strikes against her. He assumed that Lenore would not succeed in the graduate program and suggested that MIT was not a place for women—advice, as he put it, that he would give to his own daughter. It would not be surprising for a woman in this situation to have been sufficiently discouraged to choose some other path. And indeed many women who were confronted with these kinds of barriers did just that. It is hard to challenge the judgment of a department chair of a prestigious university. However, Lenore applied and was fortunate in having the support of another MIT professor, I. M. Singer. Because she was one of the best students in his course, Singer convinced the chair that she should be admitted to the graduate program. If it had not been for the help of such a mentor, it is not clear whether Lenore would have gotten in.

Like so many women who have these kinds of early experiences that cast doubt on their ability to succeed, Lenore was more determined than ever to prove she was exceptional. This unbounded enthusiasm made her bite off more than she could chew. In her first semester she tried taking eight courses (rather than the normal two or three), an impossible load. These unrealistic expectations and goals led to a difficult, discouraging, and lonely first two years. After a while, however, she adjusted to the rhythm of graduate work.

Lenore's true blossoming occurred when she started doing independent work, which culminated in her thesis. She was interested in finding ways to apply logic to other areas of mathematics, particularly algebra. During this period she developed some important ideas, which were to become the seeds of a new area of research.

When Lenore finished graduate school at MIT in 1968 at the age of twenty-five, there was every reason to believe she would have a successful and productive career as a research mathematician. She had produced an excellent thesis and had drawn the praise of very strong mentors in her field, including Gerald Sacks at MIT and Abraham Robinson at Yale. While many mathematicians doubted any woman's serious commitment to a professional career in math, assuming that once children arrived, a woman would leave mathematics and devote herself full-time to motherhood, Lenore defied these expectations. During graduate school she did indeed have a child, but she found that it in no way interfered with her professional productivity. In fact, one of her advisors was so oblivious to her pregnancy that when she arrived one day with her new baby, he looked very confused and asked where the baby had come from.

Lenore discovered in her last years of graduate school an area of mathematics that she was very excited about, applying model theory to algebra, and it was clear that there was much potential for fruitful research in the coming years. After she graduated, Lenore had job offers from some of the best schools, including MIT, Berkeley, Yale, and Purdue. Fortunately, Manuel had offers at some of the same institutions, including Yale and Berkeley. Lenore was excited about embarking on her career. This good fortune and rosy picture began to change quite dramatically, however, in the next few years.

Berkeley

After graduate school, in 1968, Lenore was awarded a one-year postdoctoral fellowship to go anywhere she chose. Berkeley seemed like an ideal place to go—there was a logic community at Berkeley, and Manuel had a job there. At the end of that year, she received an offer from MIT, as well as tenure-track offers from Yale and Purdue. When Berkeley heard about these other offers, they offered her a two-year lectureship, implying that it was comparable to her other offers; they also offered her a reduced teaching load as an extra incentive.

Lenore was naive about the differences between these different kinds of positions. And no one counseled her about the wisest choice for her career. Berkeley said that offering her a lecturer position rather than a tenure-track position allowed them to be more flexible in her pay, and so would work to her advantage. She had no reason to doubt them. Since Lenore thought that Berkeley was a better place for Manuel than Yale, and she thought it would be easier for her to adjust to Berkeley than for him to adjust to Yale, she decided to stay at Berkeley.

Two years later, the impact of that decision hit hard. Berkeley did not renew her position. Though they had assured her that they would, she had nothing down on paper. Meanwhile the job market had tightened up considerably and, having been out of graduate school for several years, she felt distant from her advisors and mentors who had helped her secure her first job.

When Lenore brought the situation, which she saw as almost a breach of contract, to the administration, they offered her an office (which she still has to this day), but did not renew her position. As Lenore describes later, she and the department had made a "gentleman's agreement" about her position. Normally this would have been honored. But since she was not a *gentleman,* they did not feel like they had to abide by it.[5]

The different expectations that Lenore and the Berkeley mathematics department had about her job and her future are reflected in how they each perceived the situation of Julia Robinson, a research mathematician affiliated with the Berkeley mathematics department. Like Lenore, Julia was married to a Berkeley faculty member. And like Lenore, Julia was not offered a tenure-track position. Julia came at a time when anti-nepotism rules were strong, and because her husband had a position in the mathematics department, she accepted her marginal status. Not until many years later, after she had been admitted to the National Academy of Sciences, was Julia given the appropriate recognition of a professorship. This appointment led to further long-overdue professional recognition, including a MacArthur Prize for her work on Hilbert's 10th Problem.[6]

When Lenore first arrived at Berkeley, she was not aware of the politics of status. Julia simply symbolized a woman who had been at Berkeley for some time, and whose husband was also at Berkeley. With this as her model, Lenore assumed that things could work out for her as well.

The department also looked at Julia's case as a model, but with a different perspective from Lenore's. They saw in Julia a resource to the department with a minimal commitment on their part. Because she was the wife of a department member, her presence was accepted, but as a woman, they did not take her completely seriously as a research mathematician. Since Lenore, too, would be around, given that her husband was on the faculty, they would also accept Lenore's presence, but they did not take her seriously as they would have a man with her credentials.

How was it that Lenore, who was so worldly and confident on one level, could be so politically naive about the career decisions she was making? Two factors contributed to that naiveté. One was her confidence. Though her confidence and natural determination had served her well growing up—and helped her maintain her vision of pursuing a life as a mathematician even in the face of adversity such as being rejected by

MIT as an undergraduate and almost again for graduate school—these same qualities worked against her at Berkeley. Her belief that she could make things work out wherever she was—in fact, that she could adjust better than her husband to any situation—led her into a kind of professional trap. Had she been less confident, she might have sought out more people's advice.

A second factor is that none of her advisors or mentors *offered* her advice on career decisions. Men are often also naive about career choices, but usually get tracked into channels that would protect them. It would be rare (certainly at the time) for a highly qualified male candidate to be offered a lecturer rather than tenure-track position. Furthermore, an advisor would be much more likely to step in and counsel his protégé if he were making a career decision that could hurt his future. But because women were still seen more as women (or wives or mothers) than as research mathematicians, this intervention was less likely to occur.

Lenore had hoped that she could establish herself in the logic community at Berkeley. Since her work was highly regarded by Sacks, Robinson, and a number of key logicians on the East Coast, she assumed that it would have the same reception at Berkeley. But it did not. She was surprised and disappointed to discover that they were not at all interested in the kind of logic she was doing.

> I was working in a field that I felt was really terrific, and when I came to Berkeley as a post-doc, I just assumed that everybody here would be interested in it, and I didn't understand the politics of why groups of people keep working in certain areas. It didn't dawn on me at first, but in truth the people at Berkeley were not interested in that field at all—not at all in model theory, and certainly not in the kind of model theory I was doing. I think it was viewed the same way as applied math was viewed in those years, as not really quite legitimate mathematics, and that logic applied to mathematics wasn't pure logic. It was the sixties, so everything had to be pure math, pure logic, etc. It is ironic to imagine that number theory or algebra would make logic impure.

Blum and the Berkeley logic group's different visions of what was considered "important" mathematics left her somewhat distant from her professional community, making it harder to establish the kind of mentor relationship that can be extremely important in one's professional development. Without such an advocate to speak on her behalf in the department, she could not survive. This kind of scenario is more likely to arise for women. A woman is more likely, for example, to follow her husband's career rather than the other way around. As a result, women's positions are often more tentative. In a normal tenure-track job, a person is hired because there is someone in the department who is a major ally and works hard to get and keep him or her in the department. With more informal positions, there is often no such advocate who really believes in one's research and will serve as one's bridge to the rest of the community.

It is important to note that up until then, jobs had simply been offered to Lenore, without her having to apply for them. As she said, "Applying for jobs was something you did when you couldn't get one." This was the standard approach to prestigious positions: advisors would contact their colleagues and informally make arrangements for an offer to be made. The assumption was that the favor would be reciprocated for the colleague's future graduates. Lenore took this for granted at the time; only later did she recognize how this system can be discriminatory against women (or any individual who is not well connected to the "mainstream society").

The Personal as Political

This early experience at Berkeley transformed Lenore from being naive about the politics for women in academia, to taking an active part in helping women navigate the difficult terrain of such a traditionally male field as mathematics. At this point, one turn of fate was to significantly affect her life for the coming years.

In 1971, three of her colleagues at Berkeley began a seminar on "Social Problems Connected with Mathematics."[7] As part of the seminar, Lenore was asked to run a colloquium on women in mathematics. This became very important because it allowed her to "take a personal thing and make it much larger." As she said, "Then I could see for myself how my situation was not just me, but part of a larger picture. I could use my own passion for my own problem, to get at this larger thing. I couldn't fight for myself at all, but I could fight for everybody else. I always saw myself as part of a larger picture—that's when you become political. You work within this larger framework."

After running this colloquium, Lenore quickly became known as the national expert on women in mathematics. It seemed a little odd to her to so suddenly be cast into this role, but it didn't take her long to fully immerse herself in it. She became involved in the newly formed national Association for Women in Mathematics, which Mary Gray helped found (and became the first president of) in the early 1970s.[8] Lenore served as the third president, from 1975 to 1978. The AWM was the first political organization for women in mathematics. It gathered statistics, monitored the participation of women at meetings and as speakers and officers of the major mathematics societies, supported women who were fighting legal battles of discrimination, and supplied a support network for the relatively isolated women in mathematics.

Lenore also became politically active at Berkeley, and was involved in many activities pertaining to women in mathematics and science. During this period she formed strong ties with other women in science, including Judy Roitman, who was then a graduate student in mathematics at Berkeley, and with whom she worked extensively in the AWM. The support she got from these women was crucial. Without that kind of supportive network, many of those women might not have stayed in mathematics. It was the first time Lenore began to see the value of deep friendships with other women. Before that, most of her close ties had been with men. She also began to appreciate the necessity of creating communities and the power of collective action. During this period she changed from focusing on herself as an individual to focusing on community.

Mills

When it was clear to Lenore that her position at Berkeley was not going anywhere, she had to decide what path to pursue next. Given the scarcity of jobs in the early 1970s, she was fortunate to find another position in the Bay Area. She was hired to teach mathematics at Mills College in Oakland, California. Though teaching at a small women's liberal-arts college did not fit Lenore's image of what she would be doing with her life, she poured herself into it with the dedication and professional focus that characterize all that she sets her mind to. Her work and contributions during this period convey a sense of her particular vision and values about teaching and administration.

When Lenore was hired at Mills, there was not a separate mathematics department. There were just a few instructors affiliated with the science division who taught whatever mathematics courses were offered. Mills did not see itself as focused on mathematics—it was, after all, an all-women's school, and mathematics was not a subject many women pursued. After one year, Lenore formed a separate mathematics department and became the chair. In her thirteen years as chair, it expanded dramatically. By 1980, more students were taking classes in mathematics and science than any other subject at Mills.[9]

Lenore's vision, as both a teacher and an administrator, has always been to find ways to fire up people's innate enthusiasm and ability. She encourages people to find what they are most interested in, and then she finds ways to empower them in those pursuits. She is not interested in channeling people into traditional paths that do not excite them, or in measuring their success by standards that are narrowly defined, such as the number of publications in prestigious academic journals. So when Lenore hired people for the mathematics department at Mills, she looked for people who were passionate about what they did. Their particular field within mathematics was of secondary importance. The faculty at Mills put a lot of energy into exciting their students with innovative projects and approaches to learning. Lenore believes strongly that this broader vision of professional activity is much more appropriate to small colleges than the traditional measures of accomplishment used at research universities, where the focus is on quantity of research publications rather than a faculty member's productivity in stimulating student engagement and enthusiasm. Indeed, whether it is computer science professors organizing slumber parties in the computer room or students doing original mathematical research, Mills has become a model of what an undergraduate institution can do to attract women to mathematics.[10]

Lenore's work at Mills was just the beginning of her national leadership in the area of women in mathematics. Lenore recognized early on that mathematics was a "critical filter" to an enormous segment of jobs in our technological society.[11] By not pursuing mathematics in school, women were locking themselves out of many of the most prestigious and high-paying professions. In order to address this problem at many different educational levels, she formed a productive partnership with Nancy Kreinberg, who was well connected to the pre-college educational community. Together they laid the groundwork for what was later to become the national Math/Science Network for girls and women. This network, which is based in the Bay Area, developed workshops, conferences, and publications whose aim was to encourage girls to pursue mathematics and science.

The project that propelled the network into existence was the first "Expanding Your Horizons" conference, held at Mills in 1975, to encourage young women to go into mathematics and science careers. Women professionals from all over the Bay Area came to talk about their lives and their work and to lead imaginative hands-on workshops in their fields. Not only was it very useful for the women students, who got to see that it was indeed possible to be a woman and to pursue science (without having to be "weird"), but it also turned out to be a terrific experience for the women speakers, who got to meet and network with other women in the community. These conferences became more and more popular every year. Today, EYH conferences are held at sites across the country, attracting many thousands of students each year.

The shift in her professional path helped Lenore recognize that there can be a kind of freedom in one's marginalized status as a woman in mathematics. For Lenore this meant proposing and experimenting with unconventional ideas. For example, there was a very strong bias in the mathematics community toward "pure" math and against "applied" mathematics, and as a result, there was little interaction between research mathematicians in universities and in industry. Lenore actively worked to form bridges between those two communities. This was an asset not only for the programs she helped organize for women in mathematics, but also for her own professional development. Often contacts in industry were tremendously helpful in supplying personal advice as she was making major career changes.

Another unconventional idea that proved to be quite fruitful was the introduction of sessions on the history of mathematics and mathematics education at the professional mathematics meetings. While others may have had similar ideas, Lenore and the AWM helped propel these ideas into motion by sponsoring talks on these topics at the annual mathematics meetings. Now these sessions are plentiful and quite popular. Repeatedly, Lenore says her ideas were not geared exclusively to women. She believes that they are beneficial for both men and women.

Lenore treated her work on women and mathematics as seriously as she would her mathematical research, and always believed in doing an extremely high-quality job. This professional approach is something both she and Nancy Kreinberg had in common; it helped to create strong programs that have survived and continued to thrive even now, many years after the founders left—a rare phenomenon for pilot programs. One of Lenore's talents that made her so successful in this work was her ability to communicate with many different audiences. In giving talks around the country, she gained the respect of people in mathematics, women's studies, business, industry, and government agencies, as well as both educators and administrators. Not only did her ability to cross these many bridges lead to the success of the above-mentioned programs, it was also to prove quite valuable in her later professional work.

Teaching

Because Lenore always saw herself as a researcher rather than a teacher, it was challenging for her to find a way to make teaching meaningful for herself. Her solution was not to focus on being an educator, but rather to focus on the mathematical ideas themselves. Lenore's philosophy, even in dealing with pre-college mathematics, was to share those mathematical ideas that she found beautiful and exciting.

During this period in the 1970s, Sheila Tobias and others were developing programs to help with "math anxiety." A number of schools introduced courses in pre-calculus, algebra, and trigonometry specifically designed to deal with math anxiety. Often these courses addressed students' psychological and social, as well as academic, difficulties with mathematics. But Lenore's philosophy was different. She believed that the best way to help women was to excite them about mathematics itself (rather than focus on their fears about mathematics), and that the best way to do that was to go straight to the most interesting ideas. She designed calculus courses that filled in inadequate backgrounds but taught the pre-calculus material (algebra and trigonometry) in context, so it was clear *why* this material was needed, and how it helped one do calculus. The design involved three lectures a week, as well as two workshops a week led by undergraduate tutors. These workshops helped establish bridges between older and younger students and created a cooperative and non-threatening atmosphere in which to learn.[12]

Furthermore, Lenore made a concerted effort in all of her courses to relate what she was teaching to topics at the forefront of mathematical research—a strategy that keeps both students and teachers interested in the material. She also found that having high expectations and goals for her students gave them something to strive for, and they would often go much further than they thought they could. Lenore's later efforts were focused on getting women into good graduate programs and inspiring them to become mathematicians—something almost inconceivable at Mills before she arrived. Several years ago, Mills initiated the highly acclaimed Summer Mathematics Institute for women from colleges around the country to give them a taste of real research and to help create a supportive community in which they can learn and do mathematics.

The overriding theme for Lenore in teaching is helping students grasp a larger vision of what they are studying, and trying to motivate them to take up mathematics on their own. She is not good at holding students' hands, and so realizes that she is not necessarily a good teacher for all students. For this reason she does not see herself as a perfect teacher. In fact, much of what guides her approach to teaching is what she describes as self-interest. She needs time to pursue her other work, so her goal as a teacher is to light a fire under her students, but not necessarily to be there the whole time it is burning. As she says, "It's like my kid. He's growing up great, but I'm not the perfect mother. And I think he's great *because* I was not the perfect mother."

Return to Research

When Lenore decided to shift gears and return to research in mathematics, letting go of her work on women in the field, many women were particularly disappointed with her decision. It was hard for them to understand how she could give up such important work that was helping so many women.

> People got really upset when it was time for a change. They thought that either I didn't value what they were doing or I was [selling out]. That was a big shock to me. I had done all this work to encourage women in science, and I thought they would cheer me on. On the other hand, I assumed that when I returned to mathematics I would have little support from the research community. Actually it was quite the opposite. The old community wasn't as supportive as the new one.

But Lenore felt that she had to return to research, both for herself and because in the end it is also important for women to have more role models of women active in research. Her move also serves as a reminder that it is possible to resume research later in one's life, though the path to doing that may be quite difficult. Some colleagues worried that she would set a bad example by staying out of mathematics for ten years and then going back in. However, it became clear to Lenore that women's lives are often different from men's. While mathematics is sometimes described as a "young man's field," women often follow a different pattern, reaching their mathematical maturity much later in life.

Because it is not unusual for women to feel somewhat outside the mainstream image of what it means to be a mathematician, it is in some ways easier for them to contemplate doing something unconventional such as returning to mathematics after being on the outskirts for ten years. At the same time, Lenore makes a point of not wanting to glamorize the experience. It is a rough road to reentry. Though you have the advantage of bypassing unnecessary work, and building on other people's work, you may have a lot of gaps in your background. In fact, one of Lenore's major goals in returning to research was to "solidify things." She compares doing mathematics to being a good athlete: you need to stay in condition if you want to perform well.

Indeed, for Lenore it was hard to reconnect to being a full-time researcher, and it took her several years to get to that point. There were no examples for her to follow, since very few people try to make the reentry into mathematics after having been out of it for a while. And no women she knew of had pursued such a path. While Joan Birman had returned to research mathematics later in life, she had done so by reentering the system in an established way—she began graduate school, albeit in her mid-thirties. Lenore, on the other hand, was not reentering the system in any conventional way: through graduate school, a post-doctoral, or a tenure-track position at a research institute. There were no well-defined steps to follow. What was clear to Lenore, however, is that she *had* to do it. "I was quite desperate. . . . I was compelled—it's sort of saving yourself from something. If it didn't work out, if I couldn't do it, I don't know how I would have accepted it. I really don't because I just so much had to do it."

There is no doubt that there are many things Lenore could have done. She has many talents, and indeed she had other possibilities open to her, but she was extremely driven to make this change and return to research. This determined drive was an important ingredient in her success.

What strategies did Lenore use to regain access to full-time research? She had become a veteran at recognizing the critical role of finding or creating a community to support one's professional development, so that is what she set out to do: reestablish herself in a research community.

Fortunately she had a friend who was a Soviet immigrant and who was a very bright and well-known computer scientist, but whose work was hard to read. She took a leave from Mills and began to travel to Boston to work with him. She thought she could begin to get back in touch with current research by helping to make his work understandable to other people. "I had no formal way of getting back in, so I was latching on to somebody that I could play a role with. . . . He needed somebody to help him with ideas, and I could get involved in that." Lenore would travel back and forth between MIT and California to meet with him. Though she didn't have a position at MIT, she had a special chair at Mills that gave her some money to travel with.

One difficulty with this arrangement was that her son was young and in school. She didn't want to move him, so she traveled for a week or two at a time, so as not to be away too long. This separation was difficult for her. It took several years of commuting before she was ready to leave for a whole semester.

During this time, Lenore became intrigued by notions of randomness, in particular from the point of view of computation. Can a computer generate sequences that behave like tosses of a fair coin? She and Manuel did some nice work investigating this seemingly paradoxical question. Yet she could not see where it was going. She needed a larger vision of her work—how it fit into a larger research plan, and how it might lead to a job.

Research

Lenore's port of entry to "a larger vision" came from two colleagues, Mike Shub and Steven Smale, who had studied randomness but from an entirely different point of view. They became interested in what Lenore was doing and encouraged her to apply for an NSF grant for women. She did apply and was awarded one. This proved to be a very significant break for her career. It allowed her to go to the Graduate Center in New York for a year and work with Shub. Though she had been treated well when she visited MIT, it was such an informal arrangement that it always felt like she was "dropping in, it wasn't legitimate." Her formal position at the Graduate Center, on the other hand, was an asset in how she thought of herself and how she was received there. It helped that community to recognize her and take her seriously as a mathematician.

Lenore credits the NSF fellowship as one of the major factors that changed her life. "That's when the work really started to take hold. . . . People started looking at me much more professionally." It was during this period that she also began to define the direction of her research. She collaborated and published several papers with Shub, a relationship that continues to be very important to her.

Finding a mathematical community, as Lenore was able to do at the Graduate Center, is a crucial ingredient to success and productivity. Such a community serves as a support group, a place to test out new ideas, and a source of constant inspiration. It is what enabled Lenore to return to research.

> There is a group of people who I am very close to, who work in my field. We are good friends. Mathematically we're very close and we're very supportive of each other. People you feel you can totally trust mathematically. You call them and they tell you what they're working on. It's a very nice feeling.

> *But these communities take a lot of work to build and sustain.*

> I've created my own institutions all along. I never had good established institutions, so I just created them. I became my own institution. It has rewards but it's also hard. If you're teaching at a place like Berkeley, there is a great institution there and you don't have to worry—it will always be there. With my way of doing it, I don't have some of those nice features; a lot of work goes into making sure it still exists year after year.

One of the major attractions of her current position at ICSI was the fact that there was an established research community that she could become a part of.

The issue of how one gains entry to and maintains a community, and the impact this has on one's mathematical development, is central to understanding the more subtle ways women continue to have difficulties being complete citizens in the arena of mathematics. Lenore had a lot of practice building communities among women at Mills, with the AWM, with women at Berkeley, with women in the Bay Area. So it seemed like a natural and doable task to create a research community. But if women do not stumble into such a research community and do not have practice creating one, their professional lives can be lonely and frustrating. It is hard to continue in isolation. A professional community gives you support, encouragement, and stimulation.

Family Life

Clearly Lenore's professional and personal lives were entwined. Her decision to take the Berkeley job over the other offers was at least in part based on the fact that Berkeley was a good place for her husband. But for the most part, Lenore was careful to keep the two spheres of her life separate—at least in public. At the time, talk of personal matters would just fuel the image of a woman as mother and not a serious mathematician. "I had spent so much of my life not having any of my personal life involved with my work. I had been married, had a baby, but we never talked about it. We wanted to make sure that we were very professional, so you would never talk about your home life."

Just as Lenore experienced professional pressure to downplay family life, she also experienced the pressure to fulfill the traditional duties associated with being a woman. Both her parents and her in-laws worried that Lenore's career would get in the way of important matters such as marriage and children. There were few role models for Lenore at that time of women who did otherwise. So, as she put it, getting married and having children early were things she needed to "get out of the way" in order to pursue her intellectual passion.

Lenore and Manuel had been married for five years when they had their first and only child. Avrim was born near the beginning of her thesis research. Lenore describes the delicate balance between her roles as mother and as mathematician in terms of colleagues' perceptions of her:

> I couldn't let people know I was pregnant because they wouldn't take me seriously—a whole academic year where basically nobody knew I was

pregnant. I was looking larger and larger. I was wearing jeans and a big jacket. The first time I wore a dress, it was a maternity dress but looked like an A-line dress, and the people commented on how dressed up I was for my exam—but that is not why I was wearing it. I had to wear this dress because I was pregnant. Most people didn't even think I was married. Here I am pregnant, married for five years, and this guy in my algebraic geometry class asked me out on a date!

For some people, having a child can lead to conflicts with professional work, particularly when the child is young and demands a lot of attention. But Lenore never felt that her son interfered with her research. Two factors played an important role: her flexible schedule during graduate school, and a supportive mother who came to care for Avrim during the crucial months when Lenore was completing her Ph.D. Sometimes Lenore or Manuel would bring Avrim in to work. They had an office in the basement of a distant building with padded walls, so it wouldn't disturb other people if he was there. In fact, when they recently returned to that office (visiting Avrim, who was a graduate student in computer science at MIT), they discovered that Avrim's drawings were still on the walls.

Manuel was actively involved in raising Avrim. As Lenore said, "Manuel was a really good co-parent and really involved from the start. I think he assumed the baby would be born and it would be a little adult person to talk to. I think it wasn't until he was five or six that he would play that role for Manuel. He was always very involved in taking care of him." Later Manuel became a close buddy and inspiring teacher for Avrim, constantly teaching mathematics and computer science in fun and interesting ways, using everyday props. For example, one morning when they had pancakes, Manuel asked him a variation of the classic Tower of Hanoi problem: If you have eight pancakes of different sizes randomly stacked in a pile, and you can use a spatula, stick it in anywhere in a stack, and flip them over, how many flips are needed to get them stacked with the largest at the bottom to the smallest on top? What if one side of each is burnt, and you want that side down? Avrim also recalls his father teaching him graph coloring problems from a very young age.

Part of the reason Avrim was easy to integrate into their lives was that many of their single friends loved to come over and be with them. In this way, he added to their friendship with other people. And they were fortunate—Avrim was a very easy child to raise. The difficulties Lenore had were with the system, not with having a child. In fact, when she decided to return to research, one of her professors at MIT said, "Ah, this is good; now that you've had and raised your child, you are ready to return to research." But Lenore resented this because the implication was that it was the child who had prevented her from doing research, which was not at all the case.

Clearly Lenore's family is very important to her, but she never saw that as a conflict with her professional life. I asked her if during any of the discouraging periods of graduate school she ever considered leaving math. She responded,

> No, that was an unthinkable thought. That would have been like total defeat. I couldn't imagine dropping out. That would be the hardest thing in my life to do. I think in some ways I was basically different from most women, because math was my life. I didn't identify with my marriage or having a child. So even when my child was born, I didn't say that I have a family so I must drop out. I almost had the family just to prove that I could have a family and keep going.

Marx once wrote that people make their own lives (and life histories), but they do so under conditions not of their own choosing. This very much captures Lenore Blum's life.

Lenore's story helps us see that at least in her case, her interest in women in mathematics does not grow out of some predisposition to fight a feminist cause, but rather out of the simple frustration of not being able to live out her dream of being a research mathematician—a dream that she ultimately was able to return to.

Lenore's story also illustrates the diverse kinds of obstacles that can get in the way of women's becoming mathematicians—policies of limited inclusion, such as those in the early days at MIT; discouragement by people in positions of power; lack of career counseling; a willingness to accept, and a naiveté about, lower-status positions; lack of a supportive research community; the absence of a strong senior advocate in one's field; and the difficulty of reentering mathematics at a later stage in one's life.

Much of the work Lenore did during her tenure at Mills involved expanding women's opportunities in order to give them access to mathematics. By returning to research herself, she continues this work in a different form, by acting as a model for such a path, and by stimulating ties between women undergraduates, women researchers, and graduate programs.

A recent event symbolized the ways her life has in some sense come full circle, as well as how much has changed in the process. She was attending a computer science conference, the focus of which spanned the research interests of Lenore, her husband, and their son, who is now also a computer scientist. She was there to hear her son give a talk—finally she had found a natural blending of her roles as mother, wife, and research mathematician. And just as she was reflecting on the fact that it was nice to be at ease being publicly seen as both a mother and a researcher, she was introduced to a woman at the conference who said, "Oh, I know you—the radical feminist lady."

Postscript

Full circles continue to emerge and intersect in Lenore's life. In 1992, she began a four-year stint as deputy director of the Mathematical Sciences Research Institute at Berkeley, one of the premier research institutes in the world. A major thrust of her work at MSRI has been to build bridges between the mathematics research community and the community at large. As an example, utilizing the skills and connections she developed in the 1970s and 1980s, Lenore instituted the popular "Conversations between Mathematics Researchers and Mathematics Teachers" held periodically at MSRI.

But the event which perhaps most poignantly brought together the many strands of her life was the Julia Robinson Celebration of Women in Mathematics held at MSRI in the summer of 1996 and commemorating the twenty-fifth anniversary of the Association for Women in Mathematics. The conference announcement stated the goals of the celebration, goals that have been close to Lenore's heart for twenty-five years:

- to showcase the recent achievements of women in mathematics
- to facilitate networking among women in various fields of mathematics
- to provide role models and offer mentoring for beginning women mathematicians
- outreach to area teachers and students

Lenore gave a talk during the celebration entitled "P = NP? and Hilbert's Nullstellensatz." In it she drew parallels between her research with Mike Shub and Steve Smale (recasting the famous open problem in complexity theory, "Does P = NP?" as a problem within mainstream algebra) and Julia Robinson's most important work (recasting a meta-mathematical problem into mainstream number theory). This work of Julia Robinson's ultimately led to the solution of Hilbert's 10th Problem. A goal and dream of the Blum, Shub, Smale work is to find a solution to the "P = NP?" problem.

Judy Roitman at age 7, 1952

Judy Roitman, 1996

Judy Roitman

(1945–)

Judy Roitman boldly defies the myth that mathematicians are narrow- and single-minded. She deftly weaves together the dominant roles that structure her life: Zen Buddhist and Jew, published poet, mathematician, and mother. Her life illustrates how difficult it is for women to consider pursuing mathematics when they are immersed in a culture that defines such a path as "unfeminine," and the circuitous route that led her from her initial love of language to the study of infinity. In this interview she reflects on the many ways that being female can influence one's experiences in mathematics.

Roitman did her graduate work at Berkeley and the University of Wisconsin, where she studied set-theoretic topology with Mary Ellen Rudin. She taught at Wellesley College for three years before joining the mathematics department at the University of Kansas, where she is now a full professor. Though she talks little about her deep commitment to both poetry and Buddhism in this interview, the two overlap and complement her work in mathematics. What all three of her passions share is an exploration of a kind of transcendent reality and the interface between abstract forms and everyday experience. At the same time, each gives expression to a different part of her self. In different ways, mathematics and Buddhism are both explorations of patterns of the mind, while poetry is an exploration not only of the mind, but of passion and emotion as well.

From Poetry to Logic

What was your childhood like?

I was a precocious reader. I had this cough constantly from the time I was about three until well into my teen years, and it was difficult to sleep at night because the cough would keep me, and everyone else, awake. By the time I was four, I was very bored with sitting up all night coughing and looking at pictures, so I asked my dad to teach me how to read, which he did. My mom read a lot to us and encouraged us to read. Every week we would go to the public library and take out ten books. By the time I was in fourth grade, I had essentially read every book in the children's section. So my mother had to petition the library to let me use the adult section.

Reading was the only thing I really had. I was a very unpleasant and obnoxious child. I was raised to think I was better than other kids. I really didn't have any friends from second grade until I got to high school and I got to be with the other weirdos! So I was a very lonely kid. And when you're a lonely kid, books are your friends.

I was in love with language. I was obsessed with language and still, to some extent, am. In school we had twenty spelling words per week. We were supposed to put five of them into five sentences. So I'd write a twenty-line poem with rhyme and meter using all twenty words. That's obsessive. But I liked it. That was my idea of a good time.

Was it in high school that you started getting interested in mathematics?

Oh, no. I didn't think about that then. I didn't like school math. It was in junior high school that I first learned about infinity and saw the proof that the real numbers are uncountable. I thought that was the greatest thing I had ever seen. That was the best. Much later, when it was clear that I was going to go into mathematics, I don't think there was a question of what field I was going to do in mathematics. But during junior high school and high school I wasn't thinking of doing math, because girls didn't do that!

Let me describe what it was like for girls when I was going to high school. My graduating class had eleven hundred students. I was in the top twenty students. Of the girls who were in the top twenty, all but two went to the Cornell University School of Home Economics, so that's clearly a very strong cultural message about what women do. When I was invited to join the science club, I saw no point, because I knew that women never made it as scientists. I didn't think of it in sociological feminist trends, as in women aren't allowed to, women are prevented from, women are excluded from—I thought of it as women are incapable of, because those were the messages that were coming from my culture.

In my high school geometry class, the teacher was preparing us for the New York State Regents' Exam. To encourage us, she announced that the year she took the exam, she got the top score. But instead of being encouraged, it confirmed for me that no matter how well a woman does formally, all she can ever hope to do is to teach high school. I was left with a complex and somewhat contradictory bunch of ideas that went something like this: that a woman can't do original work in mathematics and this is the proof of it, and that the reason that women don't do original work in mathematics . . . is not because they are not encouraged to do it, but because they are incapable genetically of doing it. I am an example, in fact, of someone who almost didn't become a mathematician. It never occurred to me until I was in my twenties to become a mathematician. So the question I always have is, How many women are there out there who are more talented than I am, for whom the same forces are operating?

Did you have a sense in high school that you were good at mathematics but you just didn't think you would ever do anything with it?

I was in the honors class, so I went through calculus in high school, but that didn't mean anything. I was always getting good grades and doing well in things like SAT scores, but I didn't take that seriously because everyone knew that I couldn't really be any good at it. It seemed almost random that I didn't always get the wrong answer.

Or that in your case it didn't really signify mathematical intelligence, it was just that you were good at test taking.

Yes, there was this thing called "the real work." We had never been given the opportunity to try the real work, and I knew immediately that I couldn't do it because I had no examples of women who did. . . . Just like music composition and conducting—those are two careers I think I would have done well in. Now that I am an adult I can say that, but when I was in high school I never would have considered it. Never.

I remember as a freshman at Oberlin College sitting around with a bunch of friends one day, and we were ticking off what women couldn't do, just deciding among ourselves. Not in anger, but just sort of "Yup, that's right. Women can't be mathematicians and physicists and conductors and composers in music, etc." We just believed it. We had a long discussion on how obscene it would be to have woman conductors. Could you imagine a woman up there? That would be so unfeminine! That's what we were doing. I didn't take math as a freshman. I'd gotten exempt from it since I had taken calculus in high school.

So once you started college at Oberlin, it didn't occur to you to take any more math?

Absolutely not. I was done with it. But then I transferred to Sarah Lawrence, which was a women's college. One of the things that drove me crazy at Oberlin was that I was the only woman who spoke out in the classes. I'd get back to the dorm and there'd be all these smart young women, and then we'd trot off to class and then I'd be the only smart young woman. This really bothered me tremendously. Because I'd grown up in this New York intellectual Jewish milieu where even if you were a garbage collector you spoke your mind. And it was this real intellectual free-for-all. Suddenly I was in this environment where the polite thing for young women was to keep their mouth shut and let the men talk. Boy, I couldn't do that. So I was going to drop out of college and become a waitress. My parents were really frantic. They eventually arranged for me to go to Sarah Lawrence. They drove from New York to Chicago with the application and stood there while I filled it out and then drove back to make the deadline, and went into debt there, and so did I, so that I could go. Sarah Lawrence was project-oriented. You had exams in some classes where it seemed reasonable, like in science and nature classes. But in humanities classes you mostly did projects or long papers, two or three a semester, and you were expected to work your little heart out over those papers. It was a lot more dignified. They had a lot more respect for the students, I thought, and I was very, very happy there.

That's much more conducive to the way you operate, with a very independent mind.

That's right. So I majored in English and threw myself into my lit courses. I also took anthropology and I took math for poets, but the main thing that happened is that they insisted that in order to take piano, I had to also simultaneously take chamber music and music composition. Here I had believed all my life that women couldn't compose, and suddenly I was composing. I remember walking in and telling my teacher, "Well, I can't do this because I am a girl"—and he just sort of looked at me and said, "Well, everyone here is a girl. So you're going to do this." And I was good. I was genuinely good.

That was surprising to you!

Yes. I couldn't write like Mozart. I'm not one of those people. But in the first year of composition, what he did was have you first write like Bach or Mozart, and then he started having you do different modern forms. And when I hit some of the more abstract modern forms, it was like I blossomed. I could have grown up and become a conductor. That was amazing. All of a sudden I felt like I could do anything I wanted to do.

The other thing that happened was that during the first week I was there, they had a student-parent tea, and the dean of women, who is one of these wonderful old characters, like out of an English mystery—you know that she's lesbian but she's not overt about it, and she has sort of the mannish tweed suits and clunky shoes and is absolutely dedicated to her students. . . . I wish I remembered her name—she was sitting at the same table with me and my mom. And my mom was a little weird, like it was a summer day and she was wearing a mink stole because it was Sarah Lawrence. The interplay between me and my mom was incredibly neurotic and crazy, and soon she became absolutely fascinated by us and said to me, "I think I'd like to talk to you. I want you to come to my office." Because she had asked me what I was going to be and I said, "A high school English teacher." And you don't go to Sarah Lawrence to become an English teacher. So she called me in and she said, "What do you want to be?" I said, "A high school English teacher." And she said, "At least become a college English teacher." In a way that's terrible, because we need good people teaching, but on the other hand she saved my life. Without her, who knows what I would have become. I would be a very bad high school English teacher. It's not the kind of thing I'm particularly good at, and she could see that. And she intervened at the right time. So at that point my goal was to be a college English teacher.

Meanwhile, I had an advisor who happened to be a math teacher. He kept telling me, "You're crazy, I want you to take more math." And I kept saying, "No, Mr. Cogan." But I took philosophy of language for him. He was a logician. Somehow or other, I had gotten a reputation as being interested in math, because I think I enjoyed the math for poets course and I talked about stuff like infinity all the time. And I took philosophy of language as a senior, partly under his influence. At that point I realized that I did not want to be an English professor, that writing was one thing and looking at literature as writers do is very different from looking at literature as English professors do, and that I was not cut out for that second thing. I was always getting into tremendous fights with my literature teachers, and I was not interested in teaching what they taught. My interests were very different. Since my interests were in language, I thought I might want to go into linguistics, but not on the level of specific languages, because in fact I'm very bad at foreign language, but rather on a more abstract level. I knew that in order to do that you would have to have more math. So at that point, after graduation, I didn't drive and I had to get away from my parents . . . so I moved out to California in 1966 because it was pretty and there were all these wonderful poets out there.

I was living as a hippie in Ashbury, and working as a secretary for the Union of American Hebrew Congregations. I was a fast worker, typing over a hundred words per minute, so they agreed that I could take courses part-time. The first summer I took calculus at Berkeley, and the second year I took second-semester calculus and baby algebra at San Francisco State. I did that two mornings per week. I still had my mind on mathematical linguistics. I didn't know anything about it, you understand, but it just sort of fit in. I needed more math courses so I quit my job and enrolled as a special student at Berkeley, my parents supporting me through that year. I didn't have much analysis and I didn't have differential geometry. Those are the two things that I was missing. And of course I didn't have statistics, but back then people didn't care about statistics. They didn't realize how important it was. I was preparing not for the math program but for the program in logic and methodology of science, so I had to take philosophy courses at the same time. That prevented me from taking all the math courses that I should have taken.

I was admitted to Berkeley's logic and methodology program, but I was also admitted to the University of Pennsylvania mathematical linguistics program with a four-year NSF Fellowship. Berkeley said, well, if you're lucky we're going to give you a teaching assistant. So Penn was a much better offer, but at that point I had started falling in love with math, so I called up Penn and said, "How much math can I take?" And they said, "Well, you can probably take an advanced algebra course." I said, "That's it?" And they said, "Well, yes, it's a busy program." So I decided I was going to Berkeley.

Giving up the four-year NSF?

Yes.

What a leap!

Yes. I entered in logic and methodology, knowing that I wanted to be in mathematics because at that point I had had it with philosophy. So I transferred into mathematics and stayed there. My first year of graduate school was a total disaster. I didn't know how to prove things. I had big gaps in my training. I had a lot of catching up to do.

Graduate School

What was it like for you at Berkeley?

The atmosphere for women at Berkeley at that time was incredibly horrible—for the women instructors as well as the students.

Can you give examples of that?

One example was another graduate school student coming up to me and saying, "How come you're at graduate school instead of being married?" A particular faculty member looking at my breasts all the time instead of looking at my face. A faculty member saying to another woman, "Mathematics is inherently masculine." People talking about how certain women instructors didn't fit in, and it was clear that what they meant was that these women were not neutered.

More sorts of things include not being taken seriously in conversations, being somewhat invisible in conversations, having passes made at you by professors and having the feeling that they were only putting up with you so that they could do that, being asked, "Well, I thought you were going to teach junior high," because I had done it over the summer to earn some money—just the general feeling that you were not being taken seriously with some overt comments. No one makes the overt comments now. There was one woman who became pregnant and she was married and they said, "Oh, well, you don't need as much money, so we'll take some of it away. We'll reduce your TA-ship."

When I took measure theory, the instructor gave an honors problem. This same instructor, by the way, would always ask for a woman to write things on the board because women had better handwriting, which meant that he had obviously never looked at my handwriting. So he gave this extra-credit problem, and there was a hint in the book, but I figured out a better way to do it. Another guy in the class used the hint in the book. The professor had him give his proof in class. I went up to the professor afterwards and said, "There's a better way of doing this." And he said, "Oh, but you followed the hint in the book and this other guy did it himself," when in fact the opposite was true. So it was that kind of assumption.

At that point you not only got grades but you also got written comments for your classes, and a very common written comment that a woman would get would be "Works very hard." In fact we were working no harder than the men, but it was just the fact that they couldn't accept that we might be just as good. I would get very outraged when I saw that, because I had, at that point, extremely bad work habits and was not working nearly as hard as I should. So it's like they were giving me credit for something that I wasn't doing and taking away credit for my natural intelligence. That really got on my nerves.

There were no women in full-time faculty positions at any of the major, major places. There was one woman instructor at Berkeley who would come to campus in a very pretty dress, change clothes to blue jeans and tee-shirts, and then put on her pretty dress to go home.

Not wanting to be seen as too feminine or too much like a woman.

Right. I remember when OB tampons came out and you could hold them in your hand, and I'd walk down the hall holding my little OB tampon and I thought, "If I open my hand and show this to anybody, the whole building is going to explode."

You mentioned earlier that you worked with other women during this time.

Yes. I had two patrons, Lenore Blum and then later I had Mary Ellen, and that definitely kept me going. I think a mentor or patron, someone who was ahead of you or encouraging you, was crucial. Because if I look at the group of women I went to graduate school with, many of them stopped being productive researchers. I don't know whether the percentage is higher than that for men. The fact is that the vast majority of people don't publish anything past their Ph.D. They rewrite their Ph.D. for a few papers for publication and then they drop off the planet. Men were dropping out, too, including some very, very good ones. But I think that the women who survived had either a male or female mentor who was really encouraging them into the post-doctorates, and in my case that was Lenore and then, at the University of Wisconsin, Mary Ellen.

I didn't work with Lenore on mathematics. Lenore was my friend, so when I was down in the dumps, she would pick me up again. That support and encouragement was very important. We were politically active together on issues pertaining to women and mathematics. Not every woman in the math department was drawn into this. There were some who avoided it. There were some who thought it was counterproductive.

When I went to the University of Wisconsin, the atmosphere was so tremendously different for women there. Women were really accepted, really felt part of it and were encouraged. I think a lot of that was because Mary Ellen Rudin was there. She was an important and respected presence, and she was a woman.

When Ken Kunen, a professor from Wisconsin, visited Berkeley, he treated me completely differently than any of the other faculty members at Berkeley had treated me. Like a person. Not like a woman. And like a woman doesn't necessarily mean sexually. It means just that I have been removed from their category. I am in the category of "other" and have to somehow be treated differently. One way or another I'm marked off, but there was none of that with Kunen. There was none of that at Wisconsin.

So you worked with Mary Ellen Rudin when you went to Wisconsin?

Yes, Mary Ellen was really the center of research, the mother of a new field of mathematics—set-theoretic topology. She was so powerful in her influence that she was attracting all these students and other mathematicians to this field. The model you saw at Berkeley of Julia Robinson was someone who continued to work but not interact with anyone that you saw; she was interacting with people elsewhere. But what you saw of

Mary Ellen was this person who was a very, very strong influence on her colleagues.

The quality of her work was so extraordinary and continues to be so extraordinary. It's a real phenomenon of nature. She did not go downhill. In some sense, she doesn't even stay level. Her stuff gets more and more complex. And that is such a wonderful thing to see. You get so much hope from seeing somebody like that. She would always go around saying, "Oh, I can't learn this." But in fact she does. Her latest paper uses diamond plus. If you're outside set theory, it's extremely hard to work with that stuff, to keep up and to know what's going on all the time. That kind of continual growth is really remarkable.

She is also a very, very warm person. I remember when I first went to Wisconsin, I called her up and her father had just died. She was on her way to the funeral. And instead of saying, "My father just died. Go away," she said, "I'm so sorry, my father just died, so I can't take care of you the way I'd like to, but I'll be back in two weeks." She just really encouraged people that way. There were a whole bunch of people who were interested students who came under her influence; there were these ongoing seminars that we'd all flock to in Wisconsin for those few years. Every vacation we had. I remember going there one Christmas vacation to visit the Israeli mathematician Shelah. It was Christmas Day and he said, "What are you doing here?" And I said, "Well, I'm Jewish, you're Jewish, I'm here. I'm on my way to some place or other."

Mathematical Experience

What is it that you love about doing mathematics?

Karen Uhlenbeck said it very well—she said it's a genderless world. You can enlarge that. It's a world in which there is nothing but the subject at hand. It's a world of no distractions. I don't want to make this sound escapist, but I have this feeling that psychically it's a place that you take refuge in. There is no poverty, there is no war, there is no sorrow, there *is* tremendous frustration—but there is frustration in everything; if you can ignore that, it's just like you're diving into this endless ocean. You're swimming around and looking at stuff, and you pick up a little rock and turn it around. All the stuff that is there to distract you everywhere else in your life isn't there.

At the same time, the idea that mathematics is not human—that's wrong. There is this book called *Mathematics: A Human Endeavor,* which was written for high school students, and it is intensely human. As far as we know, the human brain is the only brain that cares about this sort of abstract structures. It's following a certain aspect of mind out, as far as it can possibly go. It's not letting that part atrophy and be gone. This is something that we were all given before birth. At some point in the development of the fetal brain, the ability to think the kinds of things that we call mathematics arises, and this is a part of our human heritage. It's following this part out, just as music and art are part of our heritage, and some people are attracted to some things and some people are attracted to other things, and mathematics is a major human ability—it's a way of looking at things and doing things that some people get such immense pleasure from doing. I remember my kid, the other night, a couple months ago, he was lying in bed trying to go to sleep and suddenly he yelled, "When do I get my next allergy shot?" And I said, "Two and a half weeks." And there was silence for about thirty seconds and he said, "You can't have half a week. Seven doesn't split into two equal parts." Okay—that is in all of us. The idea that you look at the number seven and you suddenly ask yourself if you can have half of seven. That's the seed impulse there, looking at that question and asking about these things. I really believe that everybody has some of that in them. You may not choose to follow that out. For most people, it's destroyed in school; their natural questions and ideas [are squelched].

So for you, mathematics is really like exploring another world.

It's more like exploring the mind. I really see mathematics as a product of the human mind. I tend to see the entire world as a product of the human mind in a certain sense. The whole idea of what we see is shaped so much by our brain, that you're seeing it through all this stuff, you never see the thing in itself. Obviously, when you're doing it you're believing in a mathematical reality. Every mathematician, when they do mathematics, is a Platonist. You have to believe in these subtle cardinals in order to work with them. They're not just marks on a piece of paper. They are real to you. You can see them and turn them around and see different aspects of them. Almost like you move a physical object and you can see different aspects. When you do mathematics, obviously, you do that.

But you don't have to be a Platonist. I wouldn't want to get pinned down too much, because I'm not a philosopher and I don't think these things through rigorously, but if you were to pin me down and ask me if set theory is true, I'd have to say I don't know. I have no idea what you mean by that even. What is truth? I don't know what keeps me doing it. I like it. I enjoy it. Every now and then the frustration gets tremendous and I wonder if I should continue, and then I start doing this and I say, "God, this is fun." Even if I'm getting nowhere, just scribbling and scribbling and every day I just come in and cross out the previous day's scribble, etc., it is fun. You just can't help getting around that.

Is a necessary part of that sharing it with other people?

It is tremendous fun to talk with other people about it, but for me, the reason I need to do that is to keep from getting stuck. Otherwise I would go around and around in circles. So I need to have someone say, "No," or "What about . . . ," or "Have you heard of . . . before?" Yes, I need that. Some people don't. I think that most mathematicians do. I think there's a reason why the productivity rate at Harvard is higher than the productivity rate at Middlebury. It's because the people at Harvard talk to each other constantly.

And there's more people doing the same sort of research there.

Yes. And I need people physically present. There are people who can do it with correspondence. Peter Nyikos, my god, he writes these incredible ten- and fifteen-page letters, which he then xeroxes and sends to 5 other people! But I can't do that very well. I should call people up more, but I tend to be shy. I tend to think that this must be a really stupid question, and why should I waste anyone's time.

I think that's a problem with women. We're too shy. We tend to belittle our questions. Partly because of my own background, of not coming from a strong, good training, and from having so many gaps in that training, I tend to assume that just because I can't answer it, it doesn't mean that it's hard. So, in fact, if there is some gap in a proof, and it's a very technical gap, I tend to think that I'm just not working hard enough, when in fact that technical gap could give rise to an entire field of mathematics. Since Bill Fleissner has come, I have someone to go to immediately with a question, and if he doesn't know, then I feel more encouraged to ask further rather than feeling I'm stupid. He's just been here a year and I'm so pleased. He's great.

Gender and Collegial Interaction

Having contact with colleagues not only provides support, as you get older it also keeps your interest alive. I was starting to have trouble and then we brought in Bill, and now my mathematical life has changed incredibly because I have someone to talk to. There are people here in my field, but they are not people I could go to with rough ideas for proofs. Bill and I can go and talk to each other before we have something formalized and we're still groping. We can go and grope in front of the other one and we say,

"Oh, I don't know about that. Oh, that looks pretty good." I need that very much. So that has been very, very important for me.

And you two were graduate students together?

Yes, we go back a long ways.

So you really know each other. You don't have to go through that long process of feeling each other out and getting comfortable.

Right. I think because of what happened to me [as a woman] in graduate school, it's harder for us to trust our colleagues. I think I tend to be maybe a little pushy and strident because I had to be that way in graduate school to be heard. So I am generally not a very good colleague. I don't think I socialize correctly or work well with other people. Basically, I'm too sensitive to not being taken seriously, and I tend to simply fade away and give up. I have a terrible time when I go to Eastern Europe, which I used to do quite frequently because that's a real major place for my kind of mathematics. Many of the men there are so courtly that I just can't work with them. They're just not treating me straight.

What do you mean by courtly?

It's a kind of deference that makes you "other." So that you're not really getting down to what you need to do. It's like this little kind of dance. And again, I'm not saying that they're attracted to me or that they're trying to put a make on me; it's nothing like that. It's just that I'm "other," and they cannot talk to me with the same kind of vibrancy and casualness and immediacy with which they're talking to each other.

So it's harder to get into it. My own impression of Eastern Europe is that they don't have very many women who are productive mathematicians there, even fewer than we have here. When I do go to conferences there, the women I see are almost invariably graduate students. At one horrible conference I went to in Czechoslovakia, this guy that I went there to work with publicly declared his love for me. I had never spoken to the guy in my life. It was horrible. I can't work with him. I don't ever wish to speak with him again, and of course we've been in conferences together since then and I simply don't talk to him.

The University of Kansas

What has it been like for you at the University of Kansas?

The department has been supportive of me personally. For example, they protected me from committee work before I had tenure. They were very insistent on that. When you're a woman, as you know, they want you on all kinds of committees. My chair came to me and said, "You are not to do this." The first semester here I was asked if I wanted to go to some high-class camp for administrators in the Rockies. And I said, "I'm a researcher. When I haven't proven any theorems in five years, you call me." So they protected me from a lot of those pressures. But being the only woman was difficult at times. I came to Lawrence on January 1, 1978. For eight and a half years I was the only woman. We had women who were instructors, but the instructors don't take part in the life of the department. When a woman's issue would come up, I felt I couldn't present it as a woman's issue. So it was difficult because there was just a feeling that you represented all women and you're the only one. Then when there were two women, which there were last year, then it got to be a problem, because if we agreed with each other, they thought the women were conspiring, and if we didn't agree with each other, it was that the women were fighting. So you lose either way. Two women are worse than one. I'm really looking forward to three women because with three women, it's going to be like all of a sudden we're people again. Just like when there's one. At the same time, I think it will be easier to bring attention when there are things that affect women in the department, which there is actually very little of, but should that kind of thing occur.

This has been a positive place for me. The reason is that my department really wanted me when they hired me. It is a fairly low-key, easygoing place where people treat each other very decently. It's fairly family-oriented, so that when I did have a kid and found it difficult to do certain things schedulewise, there was, I think, more tolerance for that, even among the older guys who didn't have those problems because their wives stayed home with the kids, but there was more of an appreciation of the importance of the family. In fact, when I was single, I think they didn't quite know what to make of me. It was when I got married that I began fitting in better. I think the fact that I had this sort of patron in the background—they had Mary Ellen's word that I was worth something. They really respected her, and I think that made a tremendous difference, having that—someone that they really, really respect saying, "Yes, she's okay." And the fact that a senior member of the department had known me for a long time. I wasn't new.

So when you had to make scheduling readjustments because of having a child, they were accommodating?

Oh, yes. Like if I said, "I'm sorry, I can't teach at 4:00 P.M. ever," they'd say, "Fine, we'll give you another teaching schedule." But they tend to be fairly accommodating of everybody. They're a department that values good will. They really understand that if people don't get along, everybody is sunk. So there's a tendency to help each other out rather than stab each other in the back.

One thing that can be divisive is affirmative action issues. I feel very strongly that the way affirmative action usually operates mainly serves to alienate people. Now, this isn't always true. But so often I see this real backlash to the bureaucratic pressure. Who the hell are these people to tell us what to do! This kind of thing. I think that people need to have the illusion that they're doing this autonomously. When our department feels under the gun, it has, in fact, refused to hire—even someone who it would have hired in the normal circumstance—because they've been ordered to. It's a hard question.

Teaching

I'd like to talk about teaching and your experiences in the classroom. Does gender play a role in any way?

I notice that I dress up for my large lecture classes. A lot of students think I'm a graduate student. I make a big deal about telling the students my rank, not because of my own pride but because I want them to know that women can do it. I've noticed that Third World men, especially in the Middle East, are often extremely disrespectful of me, and I assume of women faculty in general. Sometimes even graduate students who have flunked their qualifying exams don't really take me seriously. I look a lot younger than I am, so that is a small part of it.

As far as men and women students, I certainly notice that the women students don't believe in themselves as much. They drop out of the honors track more frequently. Nationally, women don't perform as well in the Putnams, all the really competitive kinds of things. They don't seem to go for it. Women are more likely to have career aspirations which are below their abilities, although they often change their mind before they get to graduate school. But they've had some very good graduate students at KU.

One thing I've noticed is that when I send out a list every year of students who have applied for scholarships and ask for comments, the comments on the women don't seem to be as strong as the comments on the men, and then later the women will occasionally outperform the men. The men don't tend to lobby for the women students. Part of that is that the women tend to be more reticent in their classes, so you simply don't realize what you've got there. We had a case this year where there was this woman who actually ended up being *the* outstanding graduate in the school. She got a zillion-dollar fellowship from Stanford to study economics, was a math/economics major, got honors in both subjects. Math honors is very difficult. Last spring when her name was sent out for scholarships, I thought, Yes, she's an excellent student. But she ended up not getting one of the top scholarships because no one said how good this woman is, and I had never had her in a class. I don't think there was a lack of willingness to recognize her talent, it was that she kept her talent under a bushel. And it wasn't until she started working on her honors thesis that all of a sudden her honors advisor realized how good she was. She was just quietly getting all these A's in all these courses. She wasn't very aggressive in asking questions or anything like that. There was no way to know how good she was. So that was a real eye-opener for me. And when I pointed it out to people, they were very embarrassed.

I think that as a teacher I tend toward a certain kind of energy and enthusiasm which are not the sweet young woman kind of thing, which would make it harder for a student, I think, to be very disrespectful to me. I'm a loud New Yorker. So I think that is a kind of protection, my kind of style. One thing that I have noticed, as I said, in the large lecture classes, and generally with everyone except graduate students, if I don't dress up, they treat me with less respect. They treat me more casually. Part of it is just a question of looking younger than I am. Some of my male colleagues have had the same problem. I've been told that the level of decorum in my large lecture classes, in the sense of the lack of people reading newspapers and stuff, is much better than with other people. One reason is that I call on them. If I see a kid pick up his newspaper I say, "Close that newspaper." And they do! So I think I'm a pretty hard-ass disciplinarian in my classes. I will not tolerate certain behaviors that other people are shy about confronting. You know, there's always the fear that you tell someone to close the newspaper and they won't. But if you don't tell them, then they certainly won't. So I'm pretty good at confronting students.

In your classes, how do you try to communicate your excitement about mathematics?

I just do it. They look at me like I'm insane. You teach calculus for the business students and social science students and you get up there and say, "Okay, I'm going to tell you what the derivative is. The derivative is one of the great inventions of the human mind. It is one of the great contributions of European civilization to the entire planet. Before you had the derivative, the whole idea of how fast am I going right now was not precise." You get up and say all that stuff, and they look at you like she's crazy and I don't have to listen because this isn't going to be on the test.

You mentioned that when you were at Wellesley, your teaching style didn't really match with their expectations. Could you talk a little more about that?

When I have students junior level or above, or honors students, then I think I'm cheating them unless I give them some experience of what it's like to tackle a problem and figure out what the steps are. I think all of us are probably that way in the honors class. That's why it's an honors class. When it comes to graduate courses, I think I treat them with a lot of respect. I expect them to do certain things. People generally perform according to your expectations of them. I don't have the flexibility of curriculum that I had at Wellesley. At Wellesley I would do stuff like math for poets, where I would have them work on team projects and give them very little guidance. They would come to me once a week—each team— and I would talk to them and I would give them little nudges, but it was like working with graduate students, except it was a lower-level problem. Then I had them write up a team report. The opportunity to do that with undergraduates here is absolutely zilch. I think at a liberal-arts college, it might be felt that I wasn't as dedicated to teaching because I don't do some of the stuff that I consider to be not only unproductive, but in fact counterproductive. Spoon feeding. But at a place like KU nobody does that, so I think I'm sort of an ordinary teacher at KU.

The way I taught math for poets was highly nonstandard. I think that's the way to teach it, because you can give people a taste of being a mathematician with very, very easy material. I did a workshop up in Canada with a Buddhist group, where I was supposed to come in and do some fun math with them. And in forty minutes I had them, if not actually doing it, close enough so that I could tell them what it was and they could make sense of it, and they had struggled with mathematics. For example, you give people a Möbius strip, and they see that it has one edge and one side. Okay, what happens if you cut it down the middle, and if you give me the right answer I will give you the scissors. And there they are, they're stuck with it. They're mathematicians. They're trying to figure the damn thing out. Then when they're completely stuck, you draw a rectangle and you say, "Now, let's see what happens if this corner goes to this corner, this corner goes to this corner. Imagine if you cut it down the middle, what is going to happen to that edge. What is going to happen to that twist." The odds are that they won't completely get it but they will have struggled with it. They will have come up with hypotheses. They will have ideas. They will have had the experience of doing mathematics. If you have them work in teams and give them enough time, some of them will get it. That's what I would like to see people do in college math education, and that's what people don't get. They have this illusion that math is algorithms—applying other people's algorithms, and the algorithms come from the mind of God.

Why So Few Women in Mathematics?

Why do you think more women don't do mathematics?

I think the reason is very largely sociological and has to do with gender identification. The activities that go into being successful at mathematics, like the intensity of focus, the abstraction, etc., are things that are defined by our society as not feminine and very, very difficult. There are some people like Mary Ellen who seem to have been exempt from that [social conditioning]. It never occurred to them that what they were doing was unfeminine, and I think that that's true of about every successful woman in her generation. I think that in my generation it's still true about most successful women—somehow they were sheltered or blinded themselves to the social judgments that are made about those activities. But society sends extremely strong messages about which activities are masculine and which are feminine. Those associated with mathematics, except for meticulousness, which somehow gets defined as feminine as long as it's having to do with trivial matters, are defined as masculine. It's very hard to escape that, and I think that that is what Evelyn Fox Keller is really talking about; she's really exploring the images of science and our masculinization of it.

Along with that, of course, comes all the negative reactions that women get as they go along the hierarchy, as they move through the levels through school, as I and other people you have interviewed have talked about. What's interesting is that these reactions are coming at higher levels then they did before—I don't think younger women encounter them as early as we did.

Do you think that the factors that contribute to women's underrepresentation change at all as we get into college, graduate school, and professional life?

That's a good question. I think the expectations of women are lower, and I think that people tend to rise to the expectations held of them and that the act of will that it takes to go beyond what people expect of you is enormous. And very few people are capable of making it. So if the expectation is that a woman student will not be as good as a man who performs equally as well in the class, then she will tend to fulfill that expectation. There's also a very subtle question of what kinds of problems these advisors give their students. There's something else that goes on in graduate school. When it comes to the kind of independence, deep focus, and the seizing of something on your own that it takes to write a thesis, a lot of women haven't learned those skills. They are not skills that are seen as important to women. Men pick it up. Somehow they pick it up somewhere in junior or high school. I don't know exactly how. It's in the male culture. It's largely not in the female culture. The female culture is one that tells people to go easy on themselves. For women to get beyond that, to really push themselves, to really do what they're capable of doing, is sometimes very difficult. I don't want to sound like I'm blaming the victim, but I do think that there is a kind of socialization process that makes it harder for some women who clearly have talent and ability to really make themselves do what they have to do. Now, some people might say that they've made the choice not to do it, but I think in a lot of cases it's not a conscious choice, and they're still going for that degree and they still think of themselves as Ph.D. material, or maybe research-caliber. There's something that they have to learn about how you tackle problems.

Why do you think people leave at different stages: college, graduate school, even once they're in the profession?

I can really only talk about why I have at various times considered leaving mathematics, and I think basically it's that people get tired. There's something else that goes on for women, and that is that you get tired of not being accepted. Everyone gets tired of the frustration. Some frustration is inevitable in mathematics, but there is also the social frustration. Bill Thurston said something really good when he was asked by AWM for their newsletter to respond to why women don't go into mathematics more, and I think that some of this applies to why women leave. He said, "In high school, which is where you get the drive and interest, where most people become seduced by it, it's dominated by a certain group of white male nerds, who partially form this culture in order to avoid girls. So a girl cannot break into this." I'm paraphrasing, obviously. This is my memory of what he said. He said, "Nobody teaches math and science in high school. It has nothing to do with real math and science. You learn real math and science by hanging out with these nerds. These nerds will generally not let girls into their circle." Well, these nerds grow up and they still form, not necessarily the majority, but they do form a certain subculture, and they do somewhat shape the social atmosphere of the field. I think I can easily see women getting tired of putting up with it.

I feel funny saying that because it's hard to point to examples. And I'm very comfortable in the nerd subculture, because I was a girl nerd in high school. We called ourselves the snob group, by invitation only, roughly the top thirty students in my high school class. But one of those guys was a basketball player, and there was a very serious discussion about whether we could allow a jock into our circle. We were basically the kids who didn't go to the junior prom. We were the nerds. And it was a large enough school that there were enough girls around that it was possible to be a girl nerd. There were four or five girl nerds. I was one of them. I had acne and I hunched over, my hair was greasy and I didn't look real good, and my clothes didn't fit, and no one would go out with me. So I'm very comfortable in that subculture.

But I see some of my women students and it's real clear that they have certain social expectations when they're with people, and those social expectations are not met by a lot of the people in the mathematical community. You go to a math party, and sometimes there will be silence for long periods of time. I prefer this to when I go to my husband's humanities parties and everyone is trying to impress everybody else. I find mathematicians refreshingly don't do that. But on the other hand it can make people very, very uncomfortable. It's just different kinds of social expectations. I see that as a problem mostly for girls in high school and college, when they're making the decision of who they want to associate with, what kind of life they want for themselves. I see that as something that might possibly influence them not to pursue mathematics—especially if those social conventions make them think, Well, what's wrong with me? Why aren't people responding to me in a certain way? You don't realize that you're just coming from a different culture and that this culture, which has many social virtues, doesn't respond in the same way as others.

So there is the subculture issue, and then there is the simple having to put up with a lot of prejudice, sometimes expressed in well-meaning ways and sometimes not. Feeling the odd woman out. Feeling a little bit strange all the time because it's always difficult to be a counterexample, and the psychological pressure of that is something that occasionally has an effect on you.

Other difficulties: If you don't have a supportive spouse and you have to carry the burden of family and home, it would be very, very difficult to continue. On the other hand, some women that I know ran into trouble because they married senior mathematicians, and they then were perceived as spouses and that destroyed their career.

When I was in graduate school, I was becoming very active in the newly formed Association for Women in Mathematics. I was the graduate counsel representative, and then I became newsletter editor and then president. Mary Ellen was very upset that I was involved with AWM. I remember having lunch with her at a conference in Vancouver and telling her how important it was working with other women in mathematics, and she said adamantly, "If you want to help women in mathematics, you do mathematics." So it was not that women should not be encouraged in mathematics, it's that the only way to encourage women in mathematics was to do it and to do the best job you could. That was the best thing you could do for women in mathematics. I was being distracted. And of course, it did take time away from mathematics. There is no question of that. But at that point I was single and could work twenty hours per day, and that was okay. I don't really think it was a problem. I got a lot of emotional support from it, which I really needed. She didn't but I did. But I think for many women it can be a problem. There is a problem of priorities. Once you start having other obligations and you're not being sort of a nun in the Renaissance sense—where you don't necessarily have to be celibate but all you do is work—it can be a serious distraction.

I think women are taken seriously now as a matter of course. Well, no—they are and they aren't. They're taken seriously once you know who they are. When I was at the Atlanta meeting and I was at the AWM executive board, which is largely women who are fairly established in their careers, many of whom are full professors, almost all of whom have continued to publish, and are respected in substantial ways—we were all telling stories of being at conferences and not being heard, being ignored completely. Basically, when people didn't know us, we would be ignored. So there's still that hurdle that you have to overcome. Of somebody noticing you.

The assumption being that because you're a woman you can't be a serious mathematician?

Right.

Balancing Family and Work

You mentioned that you and your husband divide the child care equally. How did you work that out?

I'll tell you how it went from the beginning. When we knew that we were on the list for adoption, we arranged our schedules so we weren't teaching at the same time. One of us would be home and the other would be teaching. And we made the mistake of not hiring help for that first semester. Our son arrived on February 13. Friday. A month into the semester. We stumbled through our classes and barely taught and stumbled home. So then that fall we hired our nanny; she was from Britain, and she's a very, very wonderful, bright woman. Our son fell madly in love with her. When she went back to school, he was in daycare from ten to two, and one of us would be with him in the morning, the other in the afternoon, so we could each get in at least six straight hours of work. After that he was in school all day. Unfortunately he goes to bed late, so we have not been able to work at night. Parents whose kids go to bed at eight o'clock are lucky. There is no question that we both put much less time into our work. Our weekends are largely spent with family. My husband, because of the kind of work he does, is able to work at home, in the midst. But I can't do serious mathematics. I do things like plan classes, write exams, but I can't do research at home. On the other hand, what I found, knowing that I have these limitations on my time, is that my concentration is greater when I am doing mathematics. It's like I'm not going to waste those hours. And I find that I learn stuff much more quickly than I used to. I come up with things more quickly. You sort of respond to the constraints on you. So in fact the falloff in productivity hasn't been as great as you think, given the change in number of hours.

My husband is extremely supportive of everything in my life. I am blessed among women to have a husband like this. He pushes me to go to conferences. He encouraged me to go to a two-week poetry workshop. He is just really supportive of my work—all my different works. And not at all resentful. Like someone is taking me away from him. Or of him having to take care of Ben. My salary has always been more than my husband's because I'm in mathematics and he's in classics. He couldn't care less. That's just not important to us. He doesn't have this macho thing where he has to prove anything. That kind of thing that there is no rivalry. My success is his success, and his success is my success. There is no feeling that somebody is doing well at something and therefore putting the other one down, making the other one less important. And, in fact, he's probably more supportive of me than I am of him just because I need it more than he does. But we are both pretty supportive of each other.

Poetry is one of the things you encourage in each other.

Yes. That's what we have intellectually in common, poetry and an interest in Buddhism. Those are the intellectual things that we have in common. In fact, we met at a local Zen center about a year before we both got tenure. I feel very shy about showing my poetry to my colleagues. I'm a very serious poet. Extremely serious. As serious as anyone who teaches in an English department, who teaches poetry. The reason is that my poetry deals with aspects of life that I do not want my colleagues thinking of me in terms of—sexuality, relationships, etc. There is a certain kind of neutering that has to go on.

5

Double Jeopardy: Gender and Race

Myth: Only white males do mathematics.

If there are relatively few women mathematicians, there are even fewer black, Hispanic and Native American mathematicians.[1] Why? Some have argued that being a mathematician is a kind of luxury, one that many minorities feel they cannot afford. As one of the women interviewed for this book explains,

> I think one reason we have so few black mathematicians is that it's not something they are encouraged to go into. It's a cultural thing. There are so many things that gifted black students could obviously do that would be directly helpful to their community, and mathematics isn't one of them. So a very bright guy who has an opportunity at a first class-education who chooses to go into mathematics often feels he is not living up to his responsibilities. Black students tend to have a sufficiently strong sense of duty to their community that they become writers, doctors, lawyers, or business people, or something where they can use their talent in another way. It's a bit of a luxury to have a mathematical community—but I think it's a very important luxury and that we will have more black mathematicians when they get to the point where they feel that they can afford this luxury.

For some black and other minority students, it boils down to economic opportunity. To become a mathematician, one must invest not only in a college education but also in at least four years of graduate school. Even then, most starting-level jobs pay relatively little, certainly compared to what one makes after graduate work in medicine, law, or business. But for others, it boils down to values. Johnetta Cole, the former president of Spelman College, a predominantly black women's college, argues that a primary goal of educators should be to instill a sense of social awareness and responsibility. Mathematics, on the other hand, is often portrayed as being divorced from human, social, and political concerns—an image perpetuated not only by society at large, but even (as chapter 6 explores) by the mathematics community. For most people, it is not at all clear how being a mathematician can be used to increase social awareness or respond to social needs.

Those who want to use mathematics to help people in their communities might go on to teach mathematics in public schools. This, for example, is the recent focus of the well-known civil rights leader Bob Moses. He argues that just as literacy was the main ticket to citizenship in earlier decades, *mathematical* literacy is now an essential ingredient for full citizenship in our increasingly technological society. Consequently, he has been working to develop innovative mathematics programs to reach inner-city youth—students who often never make it past algebra, thereby compromising their future education and employment opportunities. However, while a Ph.D. can be useful, for example, by opening doors to funding agencies, it is not necessary—indeed, could even be counter-productive—to this kind of work. Thus, getting a Ph.D. in mathematics is hard to justify for many young students of color.

Nonetheless, some African American, Hispanic, and Native American men and women have gone on to become mathematicians. But how does the double dose of difference—being a woman and being a minority—affect one's life as a mathematician? The following two portraits of African American women mathematicians give us insight into how the intersection of gender and race gives rise to a particular set of issues and concerns. What drew these women to mathematics? What motivated them to go on to get a Ph.D.? How did being a woman and being black affect their educational experience? What are the advantages and disadvantages of being in minority or majority institutions? How did they integrate their research with the rest of their lives? What did teaching mathematics mean to them?

These two stories provide an interesting contrast. While Vivienne Malone-Mayes went to a predominantly black college and taught in predominantly black colleges before doing her Ph.D. work at the University of Texas and going on to teach at an almost all-white university, Fern Hunt began at a predominantly white institution for college and graduate school and went on to teach at Howard University, a predominantly black university in Washington, D.C. In both cases, the racial makeup of the institution had a profound impact on their experiences.

On the other hand, their stories are united by the fact that religion played a central role in sustaining both of them throughout their lives. One often thinks of mathematics and religion as disjoint, if not contradictory, arenas. Yet for these women, quite the opposite was true. Their lives highlight the invisible support systems that propel many white males throughout their careers, but often do not operate for women or minorities.

In both women's cases, teaching was not a routine chore that had to be dealt with in order to get on with research—it was a central part of their motivation to pursue mathematics. They wanted to convey the excitement of a life of the mind, open doors of opportunity that would otherwise be closed to minority students, instill a sense of dignity and self-esteem, and give their students tools to fight injustice. While they saw a love of mathematics as one reason to study math, it certainly was not the only reason.

The focus of this chapter is African American women. More work needs to be done to understand the complexity of how ethnic and racial backgrounds intersect with gender in the context of mathematics. These issues are particularly important given that according to some predictions, blacks and Hispanics will soon constitute 30 percent of the traditional college-age population, though they earn less than 4 percent of the Ph.D.'s awarded in mathematics each year.[2]

Vivienne Malone-Mayes

Vivienne Malone-Mayes with son, 1989

Vivienne Malone-Mayes

(1932–1995)

In the summer of 1961—during the heat of the civil rights movement, after Rosa Parks was arrested for refusing to give up her seat at the front of a bus, and when both black and white students were risking their lives for integration in Mississippi—Vivienne Malone-Mayes applied to Baylor University to do graduate work in mathematics. Though this was a path almost unheard of for black women, these were new times, and the changes were giving birth to a climate of hope.[3]

It was the first time Malone-Mayes had ever considered attending a predominantly white institution. She had grown up attending black schools in Waco, Texas, and much of her social life revolved around black churches. She received her B.A. and M.A. from Fisk University in Tennessee, and returned to teach at Paul Quinn and Bishop College, both small black colleges near Waco. But after years of trying to persuade her students to go on to get doctorates so that they could come back to teach in black colleges and help get them accredited, she was finally persuaded by her students to follow her own advice. And since there were no black state universities in Texas that offered Ph.D.'s, she applied to Baylor University because it was in her home town of Waco. She received, however, the following reply:

August 28, 1961

Dear Mrs. Mayes:

Thank you for your letter of August 24. I have discussed it with my superiors in office here but have nothing favorable to report. We have not yet taken down the racial barrier here, although I have been hopeful that it would be done eventually. It seems that everyone is waiting for everyone else and no one will take the initiative in such matters.

I sincerely wish that it were possible for me to process your application for admission to Baylor University as a student.

Very sincerely yours,

Alton B. Lee, Registrar and Director of Admissions

Undeterred, Malone-Mayes applied to and was accepted by the University of Texas at Austin instead, completing her Ph.D. in 1966.

Then, in a remarkable twist of events, Malone-Mayes was offered a professorship at Baylor University. The very same institution that four years earlier had refused to admit her as a student because of her race now hired her as the first black professor. The color barrier had been dismantled.

As Malone-Mayes was soon to discover, however, this was not the end of a struggle against discrimination; in some sense, for her, it was the beginning. The obstacles she confronted were both large and small, and it took tremendous determination to pursue this path in relative isolation as an African American, and as a woman in mathematics, at Baylor.

Different Dreams

The civil rights movement of the sixties ushered in a climate of hope and change. Defining a dream was central not only to leaders such as Martin Luther King, Jr., Malcolm X, and Ella Baker, but to many others in the black community as well. It meant shaping a vision that incorporated dignity and freedom in all aspects of life, including work, education, and relationships.

In the context of education, there were two powerful but competing visions of the future for African Americans. These two positions, articulated by W. E. B. Du Bois and Booker T. Washington, convey a great deal of history, different strategies for change, and different visions of freedom. Though these arguments began decades ago, they remain relevant today, and apply to many diverse people struggling with similar issues. These two competing visions also reveal a great deal about Vivienne Malone-Mayes's life: the dominant influences from her past, and how and why she chose her future path.

Booker T. Washington, like many people from Malone-Mayes's community, believed that education should focus on giving blacks the basic tools they needed to earn a decent living. Washington promoted vocational education. For men this meant basic blue-collar trade skills such as farming, carpentry, or plumbing. For women it meant preparation for such fields as teaching and health services.

But Du Bois had a very different vision. He felt that Booker T. Washington was playing into the stereotypes of what blacks could accomplish, a vision that was far too limited. Du Bois believed in promoting the intellect and in encouraging blacks to be the best: the best doctors, lawyers, writers, scientists, and artists. His focus was on talent, and he had high expectations. Indeed, critics of Du Bois argued that he was concerned only with the top tier, "the talented tenth," and not with the rest of the black population.

Their different philosophies were in part a product of the different communities that they grew up in. Du Bois was highly influenced by northern and educated blacks, whereas Booker T. Washington's roots were with southern blacks, many of whom, like Washington, were freed slaves—and for whom simply the freedom to have a decent living was a major accomplishment and source of pride.

Two major influences on Vivienne Malone-Mayes's life—her father, and being at Fisk—in some ways embodied these two philosophies. Her father, like Washington, had a very practical orientation—financial security and economic independence were high priorities. Fisk, on the other hand, embodied the voice of Du Bois, one of its distinguished alumni. Fisk stressed the development of the intellect and the pursuit of higher education. Dreams were big—of uplifting the race by competing on the non-material plane with the white culture. While Vivienne Malone-Mayes's early life was more clearly influenced by her father, and her later life more clearly shaped by her experiences at Fisk, both visions were important to her.

Her Father's Influence

Because Malone-Mayes's father, P. R. Malone, had such a profound influence on Vivienne's sense of herself—her dreams, her ideas about success, education, and money—it is helpful to understand some of his history.

P. R. Malone's practical orientation emerged quite early. Like his parents, who were sharecroppers, P. R. worked on a farm through most of his childhood. When he was fifteen years old, a professor visiting the town thought that P. R. had intellectual talent and educational promise. He arranged for a scholarship so that P. R. could attend Texas Central College. However, as Malone-Mayes describes, her father was not prepared to take advantage of such an opportunity—he was accustomed to hard physical work, and was too restless and undisciplined for full-time study. And although he was gifted with numbers, he was weak in other basic skills. He decided to leave school. Though in many ways he continued to educate himself throughout his life, P. R. later regretted the lost opportunity for more formal education. Because of this sense of missed opportunity, he strongly encouraged his daughter to go to college.

Rather than returning to the life of a sharecropper, P. R. got a job as a stable boy for a black doctor in Waco, Texas. Working for a black family, even a black doctor, was considered much lower status work than working for a white family. But P. R. saw it as an effective path toward economic independence. Malone-Mayes explains:

> P. R. was impressed that here was the first black family he had ever known that was not destitute, and he had faith that by observing them he, too, could learn how to make a comfortable living someday. He applied himself diligently to his duties and gradually earned Dr. Conner's trust and respect. With this respect came responsibility. Dr. Conner let P. R. collect the rent on the houses that he owned. Later he allowed him to help in the management of the properties. Through Dr. Conner he met Mr. Willis, head of the Knights of Pythias lodge. Mr. Willis took P. R. on trips to the regional lodge meetings, and allowed him to count all of the money and to keep a financial record of the meeting. P. R. was always quick with figures, and he closely observed the business transactions the two men made in buying and selling property. With their encouragement, he secured a real estate license in 1915 and opened an office of his own. Real estate was a prosperous business in the "roaring twenties."
>
> P. R. saw where he could learn something. He saw where he could emulate and imitate a lot of what others were doing, and he knew he could never do that with white people. The race barrier was very deep. His idea was to outsmart it—to get around it as much as possible.

P. R.'s work not only tapped his innate ability with numbers, it gave him the opportunity to learn many practical trades, including carpentry, electrical work, and plumbing. It also taught him a lot about money. He came to recognize that what black people considered a lot of money was really on the scale of "little penny-ante deals" compared to the deals transacted by wealthy whites—but that segregation prevented most black people from grasping the huge disparity. This early training taught him to be fiscally conservative and to recognize both the power and the illusion of money.

P. R.'s practical orientation influenced Vivienne's early career aspirations. When, as a young girl, Vivienne decided that she wanted to be a lawyer, it was her father who pointed out that the decks were stacked against black lawyers, and that most of them were in fact quite poor. As he said, "All the judges are white, so who do you think they will favor in the courtroom?" Vivienne considered other professions; at one point medicine seemed like a feasible path. But she saw the life of the doctor that her father had worked for. He was up at all hours of the night delivering babies and attending the sick—a life that seemed incompatible with having and raising children. And at the time, Vivienne had dreams of a very large family. As she says, "I would have had this great big two-story house with kids hanging out the window. I could just see this huge family. But the Lord had more sense than I had." Vivienne ended up adopting two children, who are about fifteen years apart in age.

Though P. R. was very practically oriented and fiscally conservative, his one indulgence was his daughter. She was his only child, and he loved and supported her in every way he could. In the evenings he would work with her on her homework, and while he had little formal training of his own, he had a natural understanding of mathematical ideas. He enjoyed gently coaxing her into discovering underlying patterns behind what she was doing.

> He and I would sit down, and he would listen to my spelling words and my multiplication tables. I think he enjoyed learning, too. One incident stands out in my mind. I will never forget it. I had had 5's and then we were on the 7's, and we just took them one by one. We never tried to see relationships between anything. So my daddy said, "What's after 7 times 1?" I answered, "7 times 2," "7 times 3," etc., and when we got to 7 times 5, I couldn't answer. So Daddy said, "What's 5 times 7?" And I said, "35." He said, "What's 7 times 5?" And I said, "I don't know." He sat there and I don't know how many times he had to say it before I caught on, but he never told me the answer of 7 times 5. He kept saying, "What is 5 times 7"? He taught me, without knowing, the commutative law of multiplication. That's how I know it now. No one else had ever pointed that out.

P. R.'s desire to help Vivienne on her schoolwork grew in part from his love of learning. One cannot help but wonder what direction his talents might have gone had he had access to more formal education himself.

> He could add a column of numbers in his head, going down with his finger so quickly. I was frightfully embarrassed: Here was my father, who hadn't even finished high school, and I would be scrambling around trying to add this column of numbers. Before I was halfway through, he'd be done, and have checked his work on an adding machine.
>
> He was treasurer of the church. Every Sunday he had to add up

everyone's contribution. That's where he developed his skills. He was also very quick at word problems.

I always felt that he was much smarter than I was, and if he had had the opportunities and nourishment to develop those talents, he could have done really well. But he did great despite his lack of opportunities.

P. R.'s support for his daughter was manifest in many other ways as well. Even when her actions were in discord with what he would have chosen for her, he expressed his support in clear and quiet ways. For example, many years later, in 1965, Vivienne was asked by the president of Bishop College to picket in front of a local 7-11 which was refusing to hire black employees though it was in the middle of, and making money from, the black community.

I called my father thirty minutes before going to the picket line . . . and told him what I was doing. I didn't ask his permission because I knew he would have said no. He was afraid that I would be hurt. I got out there with my sign and held it up, and when I looked up, there was my father's car. I was out there for over an hour, and every five minutes he was circling that block in his car to make sure nothing was happening to me.

P. R.'s strong base of support was to prove quite important later, when Vivienne was to confront increasing challenges in both her personal and her professional life.

Fisk

There was no question in either Vivienne's or her parents' minds that she would go to college. College was seen as a practical investment. It was an investment in Vivienne's future because it meant that she would probably get a better-paying job, one that afforded more respect than she might otherwise be able to get. But it was also an investment in Vivienne's parents' future. It was clearly conveyed by both her mother and her aunt that one of the primary purposes of attending college was to find a husband, preferably a doctor.

My father wanted me to go to Fisk because it was in the South, and they had brainwashed me to try to come back to Waco. In those days there were no rest homes and no nursing homes, and I was their only child. They didn't want me so far away that I couldn't be of any help to them. My mother's idea was that I should go wherever the best opportunity was. If I married somebody that offered to take me off, then I should go with him, and if not, bring him on back here. They felt that at Howard I would have been exposed to so many black men. My mother and my aunt made no secret of it, especially after my freshman year, that one of my objectives in going to college was to get a husband. My daddy didn't much go along with that. He thought, though, that at Howard I would have met so many blacks from the North and from the East that I would not have wanted to come back (which was probably true). None of them ever came back this way. The northern blacks always felt the South was just hell and an inferno. At a place like Nashville, where you meet blacks from the South, they wouldn't mind coming back to a southern city to live.

Vivienne was raised with a clear set of values. Financial independence and security were primary goals in life, and being a woman in no way meant financial dependence. College was important, but as a means to finding a job and a husband. Coming back to the community was a high priority. And while she did indeed fulfill these expectations, Fisk was, in the process, to change her profoundly. Her vision of a rewarding life came to encompass success not only on the material plane, but on the non-material, intellectual plane as well. She also developed a sense of responsibility not only to improve the circumstances of her immediate family, but to increase the opportunities for black people in general.

The transition to Fisk was not a simple one. To Vivienne, it seemed like a different planet. Many of the students were from better-off families, so she felt uncomfortable socially. Her academic background was weaker than most, so she felt awkward intellectually.

The difference between the values espoused at Fisk and those she had been raised with went deep. In contrast to her father's focus on economics, Malone-Mayes recalled:

> At Fisk they warned us not to be dollar-conscious because they knew we couldn't compete on the dollar level—but that we could compete on the brain level. Du Bois had gone to our school, and then gotten his Ph.D. at Harvard. We were Du Boisites, you might say; we weren't Booker T. Washingtonites. And Du Bois emphasized talent: to be the most intelligent and the most intellectual. At Fisk, we never looked forward to what we wanted to do to make money when we got out. We were liberal arts through and through. To tell you the truth, my people thought we were all crazy. When I went back to get my master's, the teachers said, "Well, I don't care if you get that master's or not; you get your teacher's certificate so that you can make some money and do something!" My aunt used to say, "If you go to work, you might as well work for money." So my people, my home, had not been intellectually oriented. It had been money-oriented. This could have been, too, why I took so long to adjust, because it was so different.

Fisk developed this intellectual orientation in many ways. It was conveyed through the speakers that were invited to campus, in what and how the students learned, and in the paths they were encouraged to follow after they graduated.

> One thing that I really felt Fisk had done to me was give me a sense that learning was an all-important thing. By my junior year, I had become a scholar. They also instilled in us a sense of pride and exposed us to many role models. They brought in speakers like Durkin Marshall, NAACP men, pacifists, Harvard people, a Canadian mathematician, to speak to us at convocation. They brought in outstanding people. And I would listen to them. When I left, I was glad I had listened, because it was those people who really made an impression on me, and I would remember them.

But it was two of Malone-Mayes's teachers at Fisk who were to have the most direct influence in shaping her future—mathematics professors Evelyn Boyd Granville and Lee Lorch. Because of their influence, she switched her major from chemistry to mathematics, and later went on to do graduate work and teach college mathematics.

> Math became more fulfilling with Evelyn Boyd Granville and Lee Lorch because they began to really let me discover math; before that, math was like rabbit-in-the-hat stuff to me. Evelyn Boyd walked in class every day: definitions, theorem, proof, then example; it was almost like a little song. A lot of the fellows in the class thought it was all worthless, they couldn't stand it. But I loved it. You had to understand it to do well on the tests; you could not reproduce it by rote. But once you began to understand it, the beauty of it began to shine through. I never liked what I call reading and writing courses—interpretive courses—because whenever I wanted to give my interpretation, it was always wrong or didn't jive with how the teacher saw it. But math is so logical.
>
> The greatest thing Evelyn Boyd Granville and Lee Lorch did for me was teach me how to read mathematics. When you opened the book, it sounded like them, like their lectures, and you could read the book. Like Lee Lorch said, "You never sit down with a math book without a pencil and paper beside you." And when it says you follow, you follow, and when it says trivial, you fill that in. They taught us how to read mathematics, and when I could do that, I knew I had made it. I couldn't have made it without that.

Evelyn Boyd Granville was one of the first black women to receive a Ph.D. in mathematics in the United States. She illustrates how important it is for women to have other women in mathematics as role models. For black women to see someone in the field who was both black and female was almost unheard of at the time, and is still extremely rare.

> We admired her [Granville]. That's another reason I became so inter-
> ested in becoming an intellectual. When she was introduced at the first
> convocation, the whole student body applauded. It shocked all of them.
> This young woman was attractive and had all of these qualities you admire:
> poise, tact, softness of voice, refinement, etc. We had enormous respect
> for her.

While having a mentor that she could identify with in these ways was
extremely powerful, Malone-Mayes's second mentor, Lee Lorch, who was
a white professor at Fisk from 1950 to 1955, also played a very important
role. He was devoted to civil rights issues; in fact, he had lost his previous
professorship at Pennsylvania State University for subletting his apart-
ment to a black family. And in 1955, when the courts ruled that segregation was unconstitutional, he had tried to enroll his daughter in the closest
school, a black school across the street from where he lived while teaching
at Fisk. His daughter was turned down by the mostly white school board,
and as Malone-Mayes recalls, "shortly thereafter he was subpoenaed by
the House Committee on Un-American Activities and subsequently was
dropped from the Fisk faculty, by a split vote of the predominantly white
board of trustees, without charges or a trial."

As Malone-Mayes said, "This man has just gone beyond the call of duty
to show respect." In a tribute she wrote for Lorch, we get a sense of his
significant influence on many of his students.

> Dr. Lorch believed that the students could *understand* the material, not
> just learn to do it. He was interested in teaching them the *why* of mathemat-
> ics in addition to the *how*. . . . The students saw that he expected them to
> learn the material, and they felt compelled to live up to his expectations.
>
> In the early fifties, the idea of encouraging blacks, and especially females,
> to prepare for academic careers was almost unheard of. Since black colleges
> were so few in number, it was not economically sound to plan on teaching
> appointments or even to pursue advanced academic degrees. In those days
> we were counseled to prepare for health professions, the ministry, or public
> school teaching, the few careers which offered an opportunity for liveli-
> hood.
>
> Of the students who were in his department for at least two of the five
> years he was at Fisk, he influenced one-fourth of them to pursue and earn
> the master's degree in pure mathematics. Moreover, one-tenth of the stu-
> dents continued to the doctorate. Each known doctoral recipient credits Lee
> Lorch as the greatest influence in his choice of career.

Malone-Mayes ended her tribute with a quote from a fellow mathematician, Charles Costley: "Probably the most important lesson I learned from
Dr. Lorch and his late wife, Grace, is that my career and success as a

mathematician should be cherished and are important—but that decency is more important than success."

Once Lorch and Granville had lit the fire, Malone-Mayes learned to love mathematics, and though she still had no idea of what exactly she was going to do with this training, she decided not only to major in it but to stay on to complete a master's degree in mathematics at Fisk.

This period at Fisk not only gave Malone-Mayes the academic background she needed to pursue mathematics, it also shaped her vision of what constitutes a meaningful life: striving for expanded educational opportunities for black people, and introducing her students to a life of the mind. After she graduated from Fisk, it seemed natural, therefore, for her to go on to teach mathematics at predominantly black colleges.

Teaching

In her first teaching position, at Paul Quinn College, and later at Bishop College, Malone-Mayes had three primary goals. The first was to support and respect her students so that they would begin to have a sense of self-worth. Immersed in a racist culture, many had little confidence either personally or academically. She believed that teachers play a powerful role in modeling the kind of respect that is essential to growth and learning. Her second goal was to give her students tools of self-empowerment. As her teachers at Fisk had done for her, she taught them to read mathematics, and how to identify and solve problems. Once they were able to do that, they did not need to rely so heavily on teachers or other authorities to use mathematics or logic in their lives. The third goal was to create a path of opportunity. In some ways, the study of mathematics is ideal for black students. A discipline with right and wrong answers makes it more difficult to discriminate against people based on race or sex, and training in mathematics leads to increased and better job opportunities.

She tried to convey that math is a tool for dealing with the world, that logical and reasoning skills should not be isolated and used only in the classroom, but should be applied to life. In politics, for example, logic can be used to argue more forcefully or to understand the fallacy in someone else's argument. She also tried to convey the basic idea of starting from simple principles to prove more complex things, to understand the fundamental relationships between things rather than just looking at and trying to memorize a mass of special cases and isolated facts.

She thought it was crucial for students to learn how to read mathematics. She often talked about wanting them to learn to "be their own doctor," to analyze the underlying problem, gather the data they need, and learn how to solve it. "My philosophy was that I do not want to be the teacher you love to date. I want to be the teacher that you respect, remember, and are grateful to for years to come. I had the same philosophy toward my youth choir in my church. I had given them something that was tangible. Something you could really use later. And you can't use isolated facts. You have to use a mode of thinking."

She also believed it was critical to allow people the room to fail. As she wrote in a column for a local newspaper and in one of her sermons, too often we present only our successes; we show only the prize fish we caught, not the one we threw back in the water. We show the theorems that have been proven, and the correct answers, not the failed attempts and dead ends. But mistakes and false starts are critical to growth and learning.

At the same time, Malone-Mayes was always a tough teacher with high expectations. And her students seemed to respond well to the challenge. As she said, "When students know that you expect it of them, they rise to the occasion because they don't have too many people expecting much." Indeed, later at Baylor, in 1985, she was voted "Outstanding Teacher of the Year." As one of her students who is now an economics professor wrote in support of the award:

> It is no exaggeration to say that Dr. Vivienne Mayes's influence on my career in research is beyond measure. I have been positively affected by a number of outstanding instructors' relationships with colleagues through-out the world. Nevertheless, I can honestly say that Dr. Mayes has had the most significant impact on my career endeavors.
>
> I will never forget my initial experiences in her class. To be perfectly frank, I was not looking forward to the experience. As I am sure you can understand, as a rather shy young man from rural east Texas, I was not looking forward to the prospects of a black female mathematics professor. However, I was persuaded by a number of my friends that this lady was really something special. I quickly became infected with the enthusiasm that Dr. Mayes brought to mathematics. Over the next two years, I would complete five classes under her direction and gain a passion for mathematics that remains to this day. Despite the obvious physical pain which she suffers, Dr. Mayes brings to the classroom a zest for her subject and a true concern for her students. She literally made the discipline come alive for me.

Malone-Mayes was pleased and proud of those students who went on to get advanced degrees. But individual success, either her own or that of a few students, was only a part of her goal. She wanted her students to go on to graduate school, get Ph.D.'s, and come back to teach in black colleges in order to get them accredited. She believed that only then would future black students have the opportunity to compete with students from white schools. However, her constant pressure on her students to pursue graduate work came right back at her. While she was still teaching at all-black Paul Quinn College, her students asked her, "Why don't you go and get your Ph.D.?"

> It had never crossed my mind. I spoke to my pastor, who had gotten his degree from Bishop, a black college in Texas, and then gone on to get his Master's of Divinity at Oberlin. He agreed that I should go to graduate school. He said, "You have too much faith in Paul Quinn; you may not even have a job there tomorrow. They may stop paying you next week because money is so tight." He said he'd been to northern schools and you don't realize what you're missing. He knew that our segregated environment was not competitive.

In a sense these words echo those of her father when he talked about black finances as "penny-ante deals" relative to those of whites. Higher education for blacks and whites was separate but far from equal.

Marriage

Through her teaching, Malone-Mayes was able to break new ground and live the values that had been instilled in her while she was at Fisk. But on another level, she was still very much controlled by tradition.

By the end of her first four years at Fisk, Vivienne had succeeded in satisfying her mother's wishes. She met and married her husband, J. J. Mayes, who was also a student at Fisk. He was older than most students. After his service in the military, the GI Bill enabled him to pursue higher education. J. J. appeared to be the ideal husband. Though he was not a doctor, he did become a dentist. He attended Meharry, a nearby black dental school, while Vivienne was getting her master's in mathematics at Fisk. When he graduated, he was showered with awards, and he later set up a practice in Waco with support from Vivienne's father. But while it looked like an ideal marriage on the outside, that was far from the reality she experienced on the inside.

Part of their difficulties arose from different attitudes about money. Vivienne had been raised by her father to be astute, independent, and an equal partner in money matters—qualities that were not able to be expressed in her marriage.

My father's idea was he never wanted me to feel that I had to go out with some fellow because I needed dinner, or that I needed a fellow to help me. He never wanted me to feel that type of pressure. But it may have been my undoing in my marriage. I never felt that my husband should support me. I always wanted us to work together, the way my mother and father had worked together. My father knew how to manage money. He always said that you should be able to give an accounting of money. But when I would ask my husband how much he made this month at the office, he would say he didn't know.

These differences over money were just a symptom of a much larger social issue: the complicated dance of finding equality in a marriage. While Vivienne agreed with her father's beliefs that an ideal marriage is one of equals, and that the wife should work and be a financial partner, she believed that her husband struggled with a sense of inferiority; rather than celebrating his wife's accomplishments, he found constant ways of disparaging her.

Their marriage lasted thirty-three years, but Malone-Mayes believed that it should have ended twenty-three years sooner. For after ten years it became clear to her that she could never please her husband, and that everything she did was wrong. Over time the relationship whittled away at her confidence in herself. She began to believe that she really was stupid and couldn't do anything right.

This lack of confidence was to make the next two major stages of her life more difficult: going to graduate school and teaching at Baylor. Nonetheless, Malone-Mayes made the decision to do these things and persevered through very difficult times. Ironically, her husband indirectly facilitated this. As she says, "In fact, it was a blessing. Without his constant criticism of me in every way, I might not have gone on to do the things I did." Whether she followed this path in an attempt to salvage a sense of herself or to prove him wrong, it took tremendous courage. Ultimately, beneath the layers of self-doubt she found underground rivers of strength derived from several sources: her father's constant support, encouragement, and respect, a strong sense of religion, and the desire to open up opportunities for future black students.

Graduate School

Just as blacks often had no idea of the material wealth available to white people, they also had no idea of the opportunities hidden from them in white educational institutions. Indeed, when Malone-Mayes applied to

Baylor, which was only a few blocks away from where she had grown up, she did not even realize that they had never accepted a black student. As she said, "It was just another world over there." Ironically, it was a good thing that she was not accepted at Baylor, since they offered only master's degrees, not doctorates, in mathematics. Instead, Malone-Mayes applied to and was accepted at the University of Texas at Austin.

No one believed that she would succeed. Indeed, there were only two black women with Ph.D.'s in mathematics in the whole country.[4] Malone-Mayes felt that she had nothing to lose. The worst that would happen was that she would fail and be back where she started. What she lacked in confidence she made up for in determination.

The first term was like jumping into northern seas in the middle of winter, a kind of shock therapy.

> I remember that the advanced calculus course was taught by a graduate student and met at 7 A.M. every morning, and not only was I the only woman, I was the only black in the class. I sat right next to the door and I was truly isolated. I had no one to talk to in the class; even though there were two other girls in the class, they avoided me like I was some sort of plague, because if you're not a white woman, you can't associate with anybody except maybe handicapped men.

There were two other black men in the program. One had been there for a year and was leaving at the end of the summer. He was one person Malone-Mayes felt that she could talk to about whether she was on the right track and could succeed. He had a lot more experience and assured her that she had what it took to make it.

But she was devastated by her first exam. She prepared for a theoretical exam, the kind she was used to from college, which tested knowledge of basic theorems, proofs, and definitions, but the exam was basically a rehashing of tricky homework-type problems. When she received a failing grade, she was completely discouraged, walked out in the hall, and cried. Despite this initial setback, however, she refused to give up. For the next test she studied problems rather than theory and received the highest grade in the class.

The isolation she experienced during that first year was very difficult. She had no one to study with—there was a group of white men who would study together, but whenever she came near, they would become silent. She also missed her daughter, who was still in Waco.[5] But while there was no one for Malone-Mayes to turn to for academic help, her philosophy was that ultimately you have to take it on your own shoulders. "You can't ask anybody to help out: no one owes you this. I always felt that way. If you flunk, well, you just took too much. I just did the best I could." Vivienne's perseverance paid off. In the end, she succeeded in her classes, getting mostly A's.

In 1964 she received an American Association of University Women Fellowship. This was an extremely important asset; in addition to the economic assistance, it provided an external affirmation of her ability, and confirmation that at least others believed in her—something that was often lacking from her own community.

> Everybody thought I was crazy. They said, "In the first place you don't *need* a Ph.D. You're black. Where are you going to work?" Every time I'd come home, and even in my class, they would ask me what I was going to do with a Ph.D. "Where are you going to work?" I felt like everything would work out. I don't know why. I never dreamed that Baylor would hire me, but I just had to keep it up. So even my family, they were just tolerating me. I had no big push from my people to complete the Ph.D. or go get it or anything. They said, "We'll help you if that's what you want to do, but we don't see why you want to do it."

Malone-Mayes's community was quite surprised when she actually finished her Ph.D. But they were even more shocked when she was offered a job at Baylor, the exclusively white institution that four years earlier had denied her admission to graduate school. "Everybody congratulated me for getting on at Baylor, but my aunt had the best response for everybody. She would say, 'Yes, it is a blessing for Baylor.'"

Baylor

While the job at Baylor was a cause for great celebration, the years to come were to prove the most difficult of Vivienne's life. It was the first time she had worked at a white institution, and the cost was quite high.

> There were 11 million blacks in this country, and we felt we had the reputation of all 11 million of them riding on our shoulders. That's why I emphasize in my classes that proof by special case is a logical fallacy, because going around trying to be proof for 11 million people is a tremendous burden. I felt like I had to show and prove that blacks can produce, they can do research, they can write papers, they can teach—I always felt that I had to prove something. As I said to a colleague, "I can't afford mediocrity."

Many black women scholars have described what they call "double jeopardy," in which black women experience the double discrimination of being black and being women in a sexist and racist society, and that it is more than just the sum of these two separate forms of isolation, but creates yet a third category of experience. As the scholar Constance Carroll says,

> There is no one with whom to share experiences and gain support, no one with whom to identify, no one on whom a Black woman can model herself. It takes a great deal of psychological strength "just to get through a day," the endless lunches and meetings in which one is always "different." The feeling is much like the exhaustion a foreigner speaking an alien tongue feels at the end of the day.[6]

While this feeling of double jeopardy is one that Malone-Mayes came to understand later in her life, it was a far cry from her earlier experiences at predominantly black colleges. Indeed, at these institutions it was typical for women to be the leaders, so she did not feel stifled in that way. Hence at Baylor, the difference in attitudes about women's academic success came as quite a shock.

> I'm from a different culture, where black women are expected to work hard and excel. But most of my students at Baylor come from a culture where if white women excel, it defeminizes them. Two of my white women students would come to my office in secret to show me their report cards with straight A's, like they had committed a crime. I sat there amazed. In black schools, my classes were predominantly women, and they were the leaders in the class.

At Baylor, the difficulty of being a black woman was ever-present. As many have expressed, it is inevitably challenging to untangle how much of that is due to race, and how much to gender.

In the beginning years at Baylor, while she had to deal with being the first black teacher on campus, she at least had the support of her chair and several deans. But when the chair changed, so did her sense of place in the department. The new chair prohibited her from teaching the more advanced courses she enjoyed, relegating her to business math and service courses. When she asked to teach summer math courses, Malone-Mayes said his response was, "I haven't even offered those to the white male faculty yet."

For many years, Malone-Mayes's salary was far below those of her peers with comparable credentials. When she had a lawyer challenge this $12,000 difference, her salary was raised $5,000, still significantly below what her peers were making.

Race was a factor in the classroom as well. Her classes were almost all white. In a population of more than 10,000 students, only 150 were black. And Malone-Mayes's style of speech, with a black southern accent and a kind of sermon-like cadence, was one that many white students resisted.

Over the years her health declined dramatically. Many of her health problems were exacerbated by the extreme stresses of her professional life. In a vicious circle, her declining health contributed to some of the professional difficulties she faced in her department. And the difficulties she faced led to stress, which exacerbated her health problems.[7] What, then, kept her going despite the problems she encountered?

Motivation and Support

A deeply entrenched belief in the mathematical community is that one should do mathematics simply because one loves it. The rewards of doing mathematics for mathematics' sake should be sufficient unto themselves. Other sources of motivation are usually seen as secondary or even misguided.

But for someone like Vivienne Malone-Mayes, while loving math was an initial driving force, it alone is not what sustained her as she faced increasingly difficult obstacles. Instead, a dominant motivator was the desire to create social change and to expand the opportunities available to future black students. There is little recognition and acknowledgment of the role of other forms of motivation, but they clearly influence how one would measure success and accomplishment. With respect to Malone-Mayes's goals, she succeeded in important ways. Not only did she herself open doors that had previously been closed to other African Americans, she also encouraged her students to pursue paths that they had not even considered possible, and gave them the tools to help them succeed.

In this context, her models and heroes were not necessarily mathematicians who created the best or the most mathematics. Instead, they were people who helped promote freedom and dignity—people such as Lee Lorch, Evelyn Boyd Granville, and Hermann Sweatt.

Hermann Marion Sweatt was a black postal employee who sued the University of Texas to be admitted to law school. The court ultimately ruled that there should be a "separate but equal" facility for him, so Texas Southern University Law School was established. But Sweatt never went there, because he eventually was accepted at the University of Texas. Unfortunately, the suit and the whole process were a harrowing ordeal. His wife ultimately left him, taking the kids. He was harassed continually, and his life was threatened. When he finally went to the University of Texas, he ended up failing. He did not believe that it was the fault of his teachers; he felt they had been fair. Vivienne says he was just a broken man by that point. But unlike some of her students, she was not compelled by his story to throw up her hands and give up; it made her incredibly grateful for Sweatt's sacrifice, for she knew that if it had not been for him, she never would have been allowed in to the University of Texas for graduate work.

She remembered the words of Martin Luther King, Jr.: If a man is in shackles all his life, you cannot expect to take them off and have him run a fair race. Vivienne Malone-Mayes wanted to be sure that her future students could run a fair race. This is what gave meaning to her life.

Not only was Malone-Mayes's motivation different from that of most other mathematicians, so too were the sources from which she derived the strength to overcome the obstacles she faced. A crucial source of support was her religious faith. Recalling her years at the University of Texas at Austin, she said,

> We have a song at our church—it's a black church—a gospel-type song, "Jesus Will Roll All Burdens Away." And in the chorus it says, "All away, all away, King Jesus will roll all your burdens away, If to him you pray, He'll open doors to you, Doors you're unable to see, Just if you pray, King Jesus will roll all burdens away." And even to this day I can hardly stand to hear that without becoming teary because I think of the doors he was opening to me while I was down there wondering why in the world I was getting all this education.

In a similar way, it was her religion that kept her going in her work at Baylor.

> I've been through illnesses, virtually dead, and I thank the Lord that there's something in my energy to pull me through. Many times I've blown my chest singing Negro spirituals. I would go to the piano and play "His Eyes on the Sparrow." The chorus is "I sing because I'm happy, I sing because I'm free, His eyes on the sparrow know that he's watching over me." In other words, if He takes care of the sparrow, He will take care of me. He will protect me. And he did. I don't want to sound like I'm going crazy or I'm a mystic or something, but you know—have you ever been sitting sometimes and thinking about something and it just seems as if something outside of you gives you that little twist that you needed to push you through?

Malone-Mayes taught in many different forums—in the classroom, from the pulpit, and through the columns she wrote for local newspapers.

But in the end the lessons she tried to convey were not too dissimilar. She wanted her students and her students' students not only to have equal access to education and jobs, but to feel that they were respected as human beings and that they had the luxury to be themselves to explore, to test ideas, to fail and to try again. To Vivienne Malone-Mayes, freedom meant the freedom to learn, the freedom to make mistakes, and the freedom to grow with dignity.

Fern Hunt

*Fern Hunt teaching at Howard
University, 1991*

Fern Hunt

(1948–)

If you ask mathematicians why they went into mathematics, you will often hear something like "It was fun; I was good at it; it seemed like a natural thing to do." But for a black woman born in 1948, there was nothing natural about going into mathematics; indeed, not a single black American woman received a Ph.D. in mathematics until 1958. So for Fern Hunt—who was born in New York City in 1948, who later became a professor at Howard University, and who is now a senior researcher at the National Institute of Standards and Technology—pursuing mathematics meant beginning a journey through unmapped territory. It was hard work, for as Fern said, "I was no one's fair-haired boy."

For the most part, Fern did not receive encouragement from the usual sources: early teachers (with one notable exception), peers, or role models. How, then, did she decide to pursue mathematics, and what enabled her to achieve a successful life as a mathematician?

For Fern, commitment to research and teaching have been intimately intertwined. Teaching gave her the opportunity to convey to her students that life is much more than "what you own, the car you drive, and where you live." She tried to share the deep excitement of an intellectual life. At the same time, even with a demanding teaching load at Howard University, it was important to her to continue her research in mathematical biology and applied mathematics. In 1993, Fern accepted a full-time position at the Applied Computation and Mathematics Division of the National Institute of Standards and Technology. There she works on mathematical problems that arise in research on the physics and chemistry of materials important to U.S. industry. She continues to have contact with undergraduate students, by lecturing at colleges and universities and working with summer students.

Fern Hunt's experiences in first predominantly white, or mixed, schools, and then later at a predominantly black university, shed light on the advantages and disadvantages of each of these environments. Fern's life is an interesting contrast to that of Vivienne Malone-Mayes, who began in all-black schools and ended up teaching at a predominantly white university.

Growing Up

Fern grew up in a housing project in the middle of Manhattan, not far from Lincoln Center. It was not long after World War II, and there was a great deal of idealism about these housing projects. They were well cared for at the time; trees and shrubs were planted throughout the project, creating a haven in the middle of the city. Her father was a postal employee, and her mother went back to work when Fern was seven, as a transcribing typist for the Welfare Department. Being in a predominantly black environment created a kind of buffer from discrimination.

From the start, Fern was very independent. She was essentially an only child for seven years, until her younger sister was born. She describes herself as "a difficult kid who resented being told what to do. I was a little spoiled, probably, . . . and I was not a terrifically outgoing person." At first she didn't care much for school, and was more absorbed with emotional and personal issues than with doing well in her classes. But at the age of seven, she changed her attitude when she realized that she was in a slow reading group. That, she decided, was unacceptable. At the time there was a tracking system in the public schools of New York City, which began as early as second grade.

> It seemed to me that I could do better than this. I was comparing myself with people in the group and already knew more than they did. It just seemed to me that they were going very slowly. So at that point I started making some extra effort. Bit by bit [I worked my way up], jumping two or three groups. I started slightly below normal in reading, and ended up reading a grade or two above by the end of the year.

From this early age, she was well aware of the inequities in the public school system. Although the schools she attended were integrated, the lower-level tracks were predominantly black, and the upper tracks were predominantly white. It was the upper-level classes that had the best teachers and offered the best education. Seeing this, Fern became motivated to do well both in her classes and on standardized exams in order to ensure that she received the highest-quality education available.

But even getting into these top classes did not erase the different treatment she received because of her skin color. Most of her teachers were white. "This often meant that some of the smaller-minded ones tended not to pay very much attention to you. You didn't get very much encouragement at all." So she made a point of doing so well academically that it was difficult to ignore her. Not until junior high school did she have a teacher that she liked, a woman mathematics teacher, Freida Denamark, of whom she says, "She treated me with respect. I think she liked me. And that was definitely different." Building up a student's self-esteem was not a focus of most teachers at that time, "and even if they did, people tended to build up the self-esteem of people of the same color." Though Fern was a bright student, most teachers did not try to encourage or really connect with her.

At this early stage of her life, Fern was much more interested in subjects she learned on her own than those she learned in school. What first excited her about science were the books she read and a Gilbert chemistry set that her mother gave her when she was recovering from an appendix operation. She was nine years old. "This was a fantastic thing for me because I would just play around and do these experiments. It's kind of strange because if you look at the cover of the chemistry set—I had the junior level—it showed two boys working. I realized that it might be a difficult career [for me], but I thought I'd worry about that later."

A mathematical spark was not lit until much later, when she started learning algebra rather than arithmetic. In elementary school, "mathematics was a subject that I hated. Arithmetic was not very interesting. I always had difficulty with bodies of knowledge made up of arbitrary rules and that is what math seemed like." But later, when she was thirteen or fourteen and began to develop an aesthetic sense for art and music, she also began to appreciate mathematics.

> I began to get more of an aesthetic sense of mathematics when I started doing algebra. I think I did so well because it was a body of knowledge where things fit together. There was some structure, and that fit with my sense and appreciation of structures. It wasn't arbitrary pieces of facts, higgledy-piggledy, but some kind of inherent relationships between facts. That began to change my attitude toward mathematics. But I still didn't see myself as a mathematician because the people around me who were good at math were boys—and white.

Throughout Fern's early life, the images and messages that surrounded her consistently implied that she would not fit in as a mathematician or a scientist—both because she was female and because she was black. How, then, did Fern come to believe that she *could* be a mathematician?

In ninth grade, Fern finally had a black teacher, Charles Wilson. That changed her life. He was a science teacher with a master's degree in chemistry from Columbia University.

> It was really a lucky thing—a chemist with an excellent education and a first-rate mind. He was the existence proof that a career in science for a black person was possible. I badly needed that at that age. He had a wonderful laboratory stocked with all kinds of things, from fetal pigs to electrical motors to Bunsen burners and chromatography tubes. He knew chemistry, physics, and biology, including biochemistry, and would help me and other students with setting up our science fair projects. Mr. Wilson used the Socratic method in his teaching, which was difficult for us at times, but in the end it taught me how to think like a scientist. He was the right person at the right time, and so he had a very profound effect on my life.

> *Would he talk to you, also, about going on in science?*

> Oh, yes. I told him my interests and ambitions. He did two very important things. The first was to tell me about the Saturday program which was available through Columbia University for kids interested in studying science. In fact, this program led to my decision to become a mathematician. The second was that he encouraged me to apply to Bronx High School of Science. If you could point to a single person who had a major influence on me, it would be Charles Wilson.

Fern was indeed accepted to the very selective Bronx High School of Science. But to her disappointment, it was not the intellectually stimulating environment she had hoped for. Bronx Science was a large school with three thousand students and a male to female ratio of two to one. Most of the students in Fern's classes were a year or two younger than she was. As she remembers it, the atmosphere was immature and repressive—there was a lock on everything, and monitors were everywhere. Competition was the norm. Though the students were quite bright, most were primarily focused on getting into Ivy League colleges and high-paying professions. Few, she felt, were "genuinely interested in ideas." And the few teachers that made an impression on her were not in mathematics or science but rather in French and history. "The key thing is not so much subject area as much as the kind of teaching approach you take with students who are going to deal with ideas in their adult careers. A teacher should foster curiosity and intellectual energy. That's a rare commodity, and young people who have it need to see examples of adults who have not lost this quality but have used it to lead a satisfying life."

Though she found high school, in general, socially and politically isolating, Fern was able to piece together a community of her own. She made friends with students in the book club, and outside of school her friends came from the Saturday Science Program and her church.

The Saturday Science Program at Columbia University that her teacher Charles Wilson had told her about provided the stimulation and scientific training that would influence the rest of her life. She began by taking two science courses, including one on astronomy. It was only when she ran out of science courses that she reluctantly agreed to take mathematics. "It didn't sound very interesting, but I did it anyway. That is definitely when I changed my basic interest from chemistry to mathematics." She discovered that she loved abstract mathematics as she learned about groups and fields—a passion that she kept alive by reading mathematics books on her own throughout high school. These included the New Mathematical Library series that the Mathematical Association of America put out for high school and early college-level audiences—books such as *Continued Fractions, Numbers, Rational and Irrational,* and *Geometric Inequalities.*

During this period, Fern read not only about mathematics, but also about biology. She was excited by Mendel's genetics work on peas, and these ideas planted the seeds that later came to fruition in her research in mathematical biology. But at that point in her life, Fern had no idea that she would end up as a mathematician; she simply could not envision herself in that role. "At that point I never really thought about what being a mathematician would be like—it was not something that I thought would be at all viable. Part of the reason was that somehow I got the idea that it was a rarefied kind of profession—that you had to be either wealthy or very, very smart, smarter than I was. . . . I really didn't see it as something that I could be doing on a day-to-day basis."

Although she could not yet imagine herself as a mathematician, Fern *was* able to imagine being an electronics engineer. She persuaded her mother to buy electrical kits so that she could do some experiments at home, most of which ended up short-circuiting the house.

What, then, did finally enable Fern to envision herself as a mathematician, to imagine that she could participate in this "rarefied kind of profession"? She remembers exactly what changed her mind. It was a booklet put out by the National Council of Teachers of Mathematics that had pictures and short biographies of several mathematicians, one of whom was Cathleen Morawetz, a mathematician at the Courant Institute, whom Fern later met when she began graduate school at NYU. It was seeing a woman that made Fern think that maybe it was possible to be a mathematician after all. Admittedly, she still had never seen or heard of any *black* mathematicians, but she was going to school during a period of great social change; in those days anything seemed possible.

College—Bryn Mawr

There was never a question in Fern or her family's minds that she would go to college, but Fern was not sure where to apply. So she sought advice from the Negro College Fund, which provided both career and college counseling. Fern was given a list of possibilities, including many Ivy League schools, but she was most interested in the seven sister colleges. She wanted to go to an all-women's college, but not one that was more like a finishing school than an academic institution. "I knew that I would have to go to the most academically demanding place because that is where I would get the best education." She chose Bryn Mawr because she was told that it was the most rigorous of the women's colleges.

Bryn Mawr was a total contrast to her high school. Where Bryn Mawr had fifteen hundred students (undergraduate *and* graduate students), Bronx Science had one thousand in her senior class alone. Bryn Mawr was all women, whereas Bronx Science not only was coed but was about two-thirds men. The graceful collegiate architecture and spacious campus of Bryn Mawr were strikingly different from the newly built modern architecture of her high school. She welcomed the new environment, and was excited about starting college.

But her undergraduate years proved to be a mixed experience.

> There were things that were very positive about the college. There were a lot of very smart people; I learned a great deal from the other students. I learned about anthropology, psychology, things that normally I wouldn't have learned about at all had I gone to a large university. I probably wouldn't have known enough to have taken those courses, and I wouldn't have met the people who would take them. Going to a large university would have been like my high school experience; I would have continued in a certain groove. But as it was, in the setting of a small college, I encountered people who had very different interests, and who were coming from a different social class and background. So that was very positive. I did a great deal of reading when I was in college—general reading. I wasn't that good a student.
>
> *They didn't reform you until graduate school.*
>
> That's right! I did work at my math courses, especially in the second half of my college career. I was very serious about that. But about my other subjects, I have to admit I was not as good as I should have been.

Though the diversity of experiences was new and exciting, Fern also experienced a sense of social isolation through much of her college years.

It is hard for her to disentangle how much of that came from being black in a predominantly white institution and how much came from being less affluent than many of her classmates. But there was another big issue as well.

> I think the biggest source of my unhappiness was the basic philosophy of the faculty at that time—their attitude was that if you had a question you wanted to investigate, you should find out what everybody else who had investigated that question said. Although I agreed, I thought it was important to start out fresh and see what *you* think first and then consult other people's work. I felt out of sync with the faculty.

This independent streak and the tendency to run counter to the prevalent attitudes in her formal education are recurring themes in Fern's life. She was influenced early on by a little book about creativity and science by A. D. Moore, whose thesis was that "creativity is one of the most important attributes that a scientist can have, and that sometimes education can really work at cross-purposes to developing a kind of freshness and independence of mind." As Fern goes on to say, "I felt that the faculty were promoting something that was quite the opposite. So I was rebelling against them."

How did this rebellion manifest itself in mathematics? "I tried to be a little bit unusual and inventive in terms of solutions to problems and things like that. I was always trying to look for the unexpected. I think most of the time I was simply wrong." The professors acknowledged, however, that while her approach was often not the most efficient one, it did display creativity.

Fern learned quickly that a creative spirit necessarily entailed a willingness to be wrong. This has always been, and continues to be, an enormous challenge for her. "Willingness to look bad, to be wrong, is very difficult. . . . Things would have happened a lot faster had I been more confident about that. It took a very, very long time, and I'm definitely better than I was, but I still have reticence. You can't really make good things happen unless you make a certain number of mistakes. And you might as well go through them right away."

Fern goes on to say that it can be particularly difficult for black people in the intellectual sphere, because "if you make a mistake, then some people are likely to say, 'Well, that just goes to prove how stupid "they" are.' That's unendurable. Nobody wants to add to the stereotype."

What allowed Fern to break through that barrier of fear of failure?

> I think that my desire to become a mathematician—well, my desire to do research, something worthwhile in some sense—overcame (eventually)

whatever reluctance I had. I never knew at any point before the end that I would indeed get a Ph.D. There never was a time when I thought, This is it! It's all downhill! Until the very, very end. But I knew that if I did not risk trying, it would be certain that I would not get a Ph.D. or become a mathematician. So it was always balancing on the one hand the certainty of failure if I did not risk, against the uncertainty of success if I did risk. So I took the uncertainty.

Thinking about a Career in Mathematics

During her undergraduate years, Fern read a book by Richard Bellman about the impact of computers on mathematics in biology. It had a formative influence on her later choices.

> The advent of computing made it possible to look at models of biological systems and analyze them. This was a new field, and he made clear it was a good field for people going into research because there were lots of problems, some unsolved for hundreds of years. There was room for people of all abilities to make contributions, so I took this to heart. I don't think I did anything about it immediately as an undergraduate, but it must have started me thinking. Eventually I did work on mathematical biology.

Throughout college, Fern wanted to pursue a career in mathematics, but she had "grave doubts" about whether she was "really good enough to do that." So she wrote a letter to one of her professors in the mathematics department, Marguerite Lehr, an algebraic geometer and someone who Fern felt was "an outstanding mathematician." In the letter Fern wrote about her interest in mathematics, and some ideas she had for research. She also wrote about her doubts in her own ability. The woman responded with a warm letter saying that Fern *should* continue, and that she was encouraged by the specific ideas for research that Fern had mentioned. The professor said, "It's often the curiosity that one has about mathematics that is a sign that you have what it takes."

This was the encouragement and inspiration that Fern needed to pursue a career in mathematics. She could have simply applied to graduate programs and assumed that if she got in, she was qualified to pursue mathematics. But what she needed was the personal encouragement of someone she admired and respected to bolster her confidence. This pattern of doubt and encouragement does not occur just once in a mathematician's life—it is a cycle that continues throughout one's career: during college, graduate school, thesis work, finding a job, getting tenure, etc. Encouragement and support are needed at all of these stages for both men and women. But for women, and especially black women, that does not come as easily.

Graduate School

Fern's choice of graduate school was influenced very much by one of her professors at Bryn Mawr, Martin Avery Snyder, who had just graduated from New York University. He said it was a good place and that she could get a scholarship to go there. Although she thought about other graduate schools, including Yale, she did not know anyone who had been there, and she worried about her chances of getting in. Courant Institute, therefore, seemed like a natural choice.

Her experiences at Courant were again mixed; there were some wonderful aspects of the program, but her path was neither easy nor fluid. Though she received her master's degree, she also received a B instead of the required A on the qualifying exam and as a result lost her fellowship and her office. At the end of her second year, she left the university. She worked for a while until she was ready to return, at which point she received a fellowship from NYU, which enabled her to finish her course work and begin her dissertation. She received an office again, and she began working as a lecturer at NYU.[8]

Graduate school was the first time Fern really began to feel a sense of camaraderie with her fellow students. The faculty were focused primarily on their own research and did not spend a lot of time interacting with students. As a result, the graduate students banded together and formed a culture of their own. Fern describes it as a "very jolly and supportive" atmosphere. Students regularly worked together, particularly in preparing for exams. They would take practice exams with each other, pretending to be the committee of examiners. Even after many of Fern's peers had graduated, they would come back to be together. Although most of the graduate students were men, there was a critical mass of women which was sufficiently large to contribute to the positive atmosphere for women. A number of her peers became lifelong friends, mentors, and future collaborators.

One important source of support for Fern during her graduate years was an organization of black graduate students called the New York Mathematics Society, later renamed the Baobab Society. It was formed to create a sense of community among budding black mathematicians who were spread out all over New York City. They met regularly for many years, holding seminars in which they would talk to each other about their research or bringing in other black mathematicians to talk to the group. An important function of the group was to support each other, "to be a sounding board for students, talk about mutual difficulties, and advise each other, especially those who were doing their theses. There were many friendships born then which have more or less sustained themselves."

Graduate school is a critical period in the development of a mathematician. This is where one forms a community that will be crucial throughout one's professional life, and this community plays a powerful role in shaping one's sense of self as a mathematician. Having this kind of community is an important source of support and camaraderie and yet is often so taken for granted that it becomes invisible. But for women, particularly women of color, such a sense of camaraderie and community is not as easily formed. When Fern found it both among her fellow graduate students and in the New York Mathematics Society for black mathematicians, she had a support network that could help her through the obstacles and difficulties one inevitably faces as a graduate student. And while there were some excellent teachers at NYU, it was really the student community that she learned the most from and that defined her experience there.

Having such a community made it much easier for Fern to imagine herself as a mathematician. She was no longer alone: as a woman, as an African American, as a young struggling mathematician. The connections she formed through these graduate student communities helped her find a thesis advisor and a job when she graduated, and helped establish her research affiliation with the National Institutes of Health.

One of Fern's fellow students heard her talk about her interest in mathematical biology, and it was he who put her in touch with the man who became her advisor, Frank Hoppensteadt. Fern describes her advisor as an unusual person because he is a mathematician and at the same time is very interested in science. Both his style and his interests dovetailed well with Fern's needs.

> He had confidence in me. And he was a strong and good enough person to say so. He didn't say it often, but he said it. It's what I probably needed to hear. I came to him with a lot of skills. I think I was twenty-seven years old when I started working with him, so I was a little older with a lot of math under my belt, so I wasn't green in that sense. Even so, he somehow seemed to be able to pace me. He knew how much to give me and how much I could do, but he did not baby me. He didn't give me any breaks when it comes right down to it. He gave me a chance and he expected me to perform, and he said I was capable of doing it, and capable of pursuing a career in mathematics. He treated me like a professional, a fellow professional.

She had found an advisor, but the process of writing a dissertation was quite challenging. She was working on difficult problems, and there were very few people working in the field of mathematical biology to turn to for help. Fortunately her advisor had a good sense of how to keep her going. If she got stuck on a problem for too long, for example, he would redirect her toward a new problem, so that she would stay mathematically active. But in her last year of graduate school, a different kind of obstacle arose. Her advisor left NYU and went to the University of Utah—not a place many black people would have chosen to go to. Nonetheless, she did follow him out there for a year to complete her thesis, taking a position in the mathematics department. To her surprise, it was a comfortable and pleasant place to be. Piecing together her solutions to problems her advisor had given her, she finished her thesis midway through the year, and came back to NYU to defend it over Christmas—a major hurdle conquered.

Howard University

While Fern was at NYU, a man named Jim Donaldson came as a post-doctoral visitor. He was the chair of the mathematics department at Howard University, and he invited her to come visit Howard, which she did. It was a good visit, and he said she should look them up when she was ready for a job. That is exactly what she did, and they hired her immediately.

Fern did not learn until much later that other universities were interested in her as well. Later, these schools said they did not understand why she did not apply, but she was given no indication that there was any reason to apply to them. Fern believes that in order for affirmative action to work, schools must take a more active role in seeking out women and minorities and encouraging them to apply for jobs, just as Jim Donaldson did to lure her to Howard. At the time, there was nothing obvious or natural about applying to a host of unknown schools.

As it turned out, being at Howard was good for Fern in many ways, especially because of the tremendous support from Donaldson.

> After I finished my dissertation, I had some surgery which should have been fairly routine, but the upshot of it was that I was allergic to the anesthetic and I got toxic hepatitis. The stress of finishing the dissertation and the illness was like fighting and squeezing my way out a very narrow, tight opening. It left me tired and a little uncertain about the future and whether I would be able to do any research. I did start working again, but it was Jim who always had a high opinion of me, and I appreciated that.

He would say, "Don't worry, you'll find a problem and you'll do it!" He just took that for granted.

And indeed, Fern has managed to continue to be productive in her research even with a fairly heavy teaching load at Howard. She had leaves to pursue research at the National Institutes of Health and later at the National Institute of Standards and Technology. She attributes her productivity both to the fact that she has very few family demands, and to the fact that she has had so much encouragement in her research from people such as Jim Donaldson, Adeniran Adeboye, Tepper Gill, Gerald Chachere, and many others on the faculty at Howard University.

The Role of Religion

Like every mathematician, Fern has her good periods and her discouraging periods. Mathematics can be hard, time-consuming, and frustrating. "Sometimes I go through long periods where nothing really works out." What keeps her going during these times is her love of mathematics and her religious faith.

Many scientists who have a spiritual dimension to their lives are reluctant to talk about it publicly, because modern-day science is usually pitted against religion as if the two were contradictory. This is a relatively modern view, and many individual scientists find the blending of science and religion natural and important.

Though Fern was raised as a Christian and actively participated in church, during her late high school years she found herself becoming more disillusioned, and she finally became an atheist. In college she met what she describes as "a couple of the very few people on campus who not only went to church but really believed."

> They didn't proselytize, and in fact you had to know them well before you figured out they did go to church. They were full of jokes and puns—they were quite irreverent and at times mischievous. They gracefully blended these qualities with courage and integrity. I thought this was so odd; I knew very few people like this. Since I found the whole subject of religion to be embarrassing and beneath notice, I tried at first to overlook this weakness in them. But they were good friends to me and I really liked them, so after a while I started thinking.

Then the summer of her sophomore year, Fern had an experience that fundamentally changed her life. With the help of the dean of the college, she was offered a summer job as an engineering assistant in New London, Connecticut. The dean tried to set up housing for Fern, but it fell through. So when she arrived in New London, she tried to find housing on her own. But no one would rent her a place to live because of her skin color. It was the first time she had encountered overt and direct discrimination. She was eighteen years old.

It was not subtle and it sort of sent me reeling. I got pretty close to the edge then. I was by myself. My folks were in New York. There wasn't anybody I could go to. Somehow I needed help to survive this. It was shattering. For the first time, I understood that religious faith can sustain you. It is the real ground on which we stand—not our family and not our friends, however much they love us. I wish there were less painful ways of finding this out. As it was, I am grateful that I was "brought up short" on this issue early in life. After many years, I can say that the most important gift that religious faith has given me is gratitude for being created the way I was. Perhaps I wouldn't have been happier, and I almost certainly wouldn't have been a better person, had I been born with qualities that some view as more "acceptable."

Did you know that discrimination existed when you were growing up in New York City?

Oh, yes, I knew that discrimination was around and I saw it, but I never understood the extent of it. At home, I could always go back and tell my family about it. I had this matrix of support. During high school you saw discrimination, but also, at that time in New York City, there were still lots of people of good will who were committed to integration of some kind, and the social fabric had not yet pulled apart and the economic conditions had not worsened to the point where there was really overt racism. In some parts of New York City, things were bad, but you just avoided those places. In a completely new environment, though, I had no guideposts—college, after all, had been rather tolerant. Now there was nobody I could go to, and I couldn't find a place to live.

Total isolation.

Yes, totally isolated. I was living in a place where no one would speak to me. I was it. My choices were to turn around and go back on the train or to survive. And I couldn't go home. Maybe that was sort of foolish, but I felt I couldn't go back to New York. I needed the money and this job.

So you stayed?

Yes. And I eventually got acceptable housing. I think I called the NAACP, and they referred me to somebody. I think that's eventually how it worked itself out. But even at work the people weren't terribly forward. I have a fairly thick skin; that, plus being at home, always provided this cushion, but suddenly it was just stripped.

When you later experienced discrimination, was it less difficult because you had already been through that?

I learned to take a long view. A couple of decades down the road, these things are going to seem trivial. You have to think about the end of your life. If you sum up your life, what do you want to say? When the curtain comes down, what do you want the review to be? So several things helped me: I took a longer view, I had inspirational resources, and I also have a very stubborn personality that says, Well, I'll show them!

For Fern, dealing with doubt and uncertainty was an integral part of her life. It brought together questions of who she was as a person and as a mathematician.

The crisis of youth is the crisis of self-definition. Even in a society where an individual's role is defined early and very clearly, there is still the uncertainty of not knowing at a fundamental, spiritual level why we are here and what our ultimate purpose in life is. In a society as fluid as ours— where small differences in our choices can have large consequences—the question of definition is that much sharper. Problems of race and gender discrimination posed challenges to who I was at a fundamental level. I was forced to search for answers at that level. I think this happens to anyone who has experienced suffering over a long period of time.

Given how important religious faith is in Fern's life, how does she blend her mathematics and religion? How does she reconcile what some see as contradictory enterprises? For Fern, the belief in realities beyond our concrete physical world is what unites the mathematical and spiritual components of her life.

As a mathematician you are dealing with abstract structures, and the kinds of things that impressed me most were not really visible. So as a mathematician it is easier to believe that the kind of concrete things that we deal with every day do not constitute all of reality; there are realities that are not immediately perceptible by us. That's the basic difference between someone who is some kind of a theist and somebody who is not: the belief that reality is just far wider than simple physical laws, economic and political laws, or relationships that we see.

Lessons Learned from Teaching

Fern was motivated to teach by a desire to convey to the next generation life lessons and inspiration. "I have learned some things. I've read a lot. I have insights. I would like for the next generation to know about them. As a human being, there really isn't any better purpose for anything you do than to try to give other people the benefit of what you've learned, so that it's a little bit easier for them to advance. Teaching gives me an opportunity to do that."

In giving talks at a variety of schools around the country, Fern always finds ways to describe how interesting mathematics can be. Often the students ask questions like "What is it like to be a scientist?" Fern responds: "I tell them that it brings me a great deal of personal satisfaction. Many people feel a lack of direction that comes from not having a strong vocation or a sense of calling to do something. Income, position, being able to support a family at a high level can be motivation for a career." But for Fern, there is a very deep level of satisfaction in a life of the mind. For this reason she tries to convey that mathematics is about "how to deal with ideas and put together ideas that have some structure and to have an appreciation for them. The world is bigger than what job you have, your income, what car you're driving, or where you live—that's something you try to hold out to people of all ages, but especially younger people."

There are three life lessons in particular that she hopes to convey. The first is about gaining confidence. "I would like to convey to students a little more confidence in themselves and their ability to be able to deal with difficult problems, to know that, at least to some extent, some sense of self-mastery is to be gained by studying."

Second, she hopes to communicate the pleasure of intellectual activity. It's really quite a great life that way. It involves "coming up with the unexpected, enjoying challenges, and having a feeling and appreciation for structure." She goes on to say, "It shouldn't just be left to artists. It's something that other people should have, in particular mathematicians."

The third lesson she tries to impart is that blacks can and should do mathematics. But in order for a teacher to convey that to students, the teacher has to believe in them. She cites the Helen Keller story to illustrate that "in order to be able to teach a person, you have to take them seriously. You can't have contempt for them."

In general, "I hope to convey a certain amount of independence of mind. That is the single most important thing, that when I'm not there, they will be able to approach a problem and be able to think about it. It's a habit of mind, to be able to think about a problem, come up with a hypothesis, and be able to move toward a solution." She tries to communicate how mathematics is not in some distant world, but rather "pops up right in front of you." But the problems are disguised, so you have to recognize them in order to work on them. Otherwise "you're helpless in that situation."

Encouraging More Women and
Minorities in Mathematics

Fern believes strongly in making the field more inclusive—that one's sex, race, or class should not prevent one from studying or pursuing mathematics. She uses an analogy to basketball to emphasize that there are many levels of talent.

> Math is a lot like sports. There is a lot of talent and many kinds of talent. Take basketball as an analogy. We all appreciate players like Larry Bird, Bill Russell, Bill Walton, as well as Nate (Tiny) Archibald or Isiah Thomas. They have (had) very different abilities and styles of play. Yet they were all marvelous players. So the pool of talent is broad. And when you consider Patrick Ewing, Magic Johnson, Julius (Dr. J) Erving, and finally Michael Jordan, you also see an almost infinite depth of talent. There is probably a broad consensus that Jordan is the greatest player the game has produced, but does anyone think that diminishes the contributions of Julius Erving? Would Jordan have achieved as much without the help of the less talented Scottie Pippen? Mathematics is like that. No matter how good you actually are, there is definitely somebody who can run rings around you. If you encounter these people, it can be intimidating. This and the fact that mathematics is a field that a lot of people have trouble with causes a great deal of anxiety both within and outside the profession. I think we should minimize it by trying to be a little more inclusive, by trying to look for the talent people *have,* rather than dismissing them for the talent they lack.

Encouragement is crucial in developing mathematical talent. She gives an example of a student in one of her courses who was doing nothing; the highest grade he got on an exam was a 70. But she noticed that he was going in regularly to see one of her colleagues. Her colleague said, "Yeah, he's a lazy guy, but he is bright," and therefore was giving him some hard problems to work on on his own. The student was in fact doing very well on them, so Fern decided to follow suit, and had the student do an independent project. Not only did this engage him in that work, it improved his participation and performance in class as well.

> It's important especially when you're trying to increase the pool of so-called underachievers [to find ways to encourage them]. There are a lot of reasons for poor performance. There are more reasons for poor performance than there are for good performance. It makes things much, much more complex.
>
> *Like they might just be unwilling to speak up in class or . . .*
>
> Yes, or they feel that they can't speak up in class. It may be that they don't

have good study habits or that they don't know how to work with other students or that other students refuse to work with them. There could be other reasons, subtle and difficult ones having to do with personal issues like lack of maturity or family conflicts.

Even in her own undergraduate institution, she was disturbed by the attitude of the mathematics department, which focused only on those few students who were going on to become research mathematicians, and ignored the rest. She sees the same phenomenon in graduate programs that turn away talented graduates of small liberal-arts schools because their background is not as comprehensive as that of students coming from large, prestigious universities. This is especially disturbing because some of the small liberal-arts schools have produced a disproportionately large number of female scientists. If top graduate programs close their doors to these students, they are turning away a significant pool of potential women and minority scientists.

Ultimately, Fern believes that the most important quality for success in mathematics is intellectual curiosity. Clearly a small school cannot offer the range of courses a university offers, but Fern believes that "if a major can graduate with a solid understanding of advanced calculus and abstract algebra, I think you've got a very good product there, and I think that is somebody you can definitely work with." She argues that various mechanisms or programs could be developed to ease the transition of students from four-year colleges to graduate programs.

In summary, Fern says, "Our methods don't pick up every possible aspect of the kinds of talents that could potentially be useful. And it's for that reason that we should be more inclusive. Our tools for discerning talent are blunt and imperfect."

Recommendations

Fern has many ideas for ways to make mathematics more enticing for women and minorities. The following is a list of general ideas that can be incorporated into a classroom, followed by a list of more specific techniques for encouraging women and minorities.

PRESENT MATHEMATICS AS A HUMAN ENDEAVOR

Fern argues that "mathematics history is not taught at all, and I think that's a mistake—math history was something that I had read in high school, and without doubt it helped me make the decision to become a mathematician." She talks about how inspiring it was to read about the lives of mathematicians: "It makes the kinds of work they were doing exciting—something really worth aspiring to." By being presented in the context of its history, the subject matter becomes more human, thereby making it more appealing to a wider range of students.

A second way to reveal the human dimension of mathematics is to talk about the people behind the mathematics. Fern makes a point of talking about David Blackwell, who developed some of the material they cover in her classes, and who also was chairman of the mathematics department at Howard for ten years. Since he is black, she thinks it is especially important for the students to hear more about him than his name.

A third way to present mathematics as a human activity is to try to convey the interconnections between a society and mathematics. This is a topic that Fern finds fascinating and continues to learn about. When we spoke, for example, she had recently been reading about the French Revolution and how important it was in the development of mathematics and technology in France.

Use Real-Life Examples

Since Fern often teaches probability and statistics, she tries to find examples that could emerge from everyday life. Medicine abounds with basic statistics and probability problems. She might present, for example, a hypothetical scenario. Suppose a doctor decides to give a transfusion to a patient who is a hemophiliac. There are two pools of blood in this country, one obtained from volunteers, and the other obtained from those who donate blood for money. The latter is much more likely to have diseases such as hepatitis. The doctor might decide to use the second blood supply, which is normally used only in extreme emergencies, arguing that the probability that the patient has already been exposed to hepatitis is 100 percent for the following reason: If the patient is a hemophiliac and has already had 100 transfusions, and the probability that a transfusion contains hepatitis is 1 percent, he might multiply $100 \times .01$ and get a probability of 1, i.e., a 100 percent likelihood that the patient has already been exposed to hepatitis. But this would be a drastic miscalculation. The real probability is $1 - [.99 \text{ to the 100th power}] = 63$ percent. This kind of mathematical understanding is critical to medicine as well as many other types of professions, including law and business. Whenever possible, therefore, Fern talks about her own work, why it is important, and how it relates to the real world.

Ultimately Fern sees mathematics as a deeply human activity that gives us a broader vision of ourselves and helps us find meaning in our lives. At the same time, she sees it as a luxury that must be appreciated and passed on, or it could be lost.

> The point is that as soon as there was enough to eat and the environment was relatively stable, humans were involved in mathematical activity. But this kind of broad vision of mathematics is important in order to secure its future. There is nothing that says that mathematics ought to continue as a cultural activity. Indeed, there are many arts and crafts that have been lost as the civilizations that invented them declined. The first six books of Euclid's Elements were lost in the fire at the Alexandria library—a significant blow to the development of mathematics. It would be terrible to think of something like that happening now, but it could. It is better to invest in as many people as possible—to convey the idea that science and mathematics elevate the mind and the spirit, and that is something that people will always need. Somehow that's not being transmitted. Somehow we're letting our machines, our greed, and our own spiritual emptiness devour everything. So the enterprise of teaching, and research in mathematics— whether in academia or out—spreading the knowledge we gain to as many as possible is not only a noble activity, it is also the conservative thing to do.

The specific techniques Fern uses to create an atmosphere of inclusivity in the classroom include the following:

EXTENSIVE USE OF QUESTIONS

Most of Fern's classes involve a kind of call-and-response pattern. She makes a point of always starting with simple questions to get the students initially engaged. Once they feel comfortable thinking and talking out loud, they are more willing to try ideas they are less sure of, and are more willing to be wrong in class. If the answer is even remotely right, she tries to work with it.

Since the males in her classes tend to speak more and have more confidence, she makes an extra effort to involve women and hear what they have to say. She finds that this little bit of extra attention is quite effective in drawing women out.

ENCOURAGEMENT

Fern believes that the most important thing of all is to encourage students. Everything else pales in comparison. This is especially important with any student who expresses interest in mathematics; and since some may express it in very quiet ways, it involves paying careful attention. For example, a student might attend math club meetings but never speak up, or may be very attentive in class but shy about answering questions. So how does one encourage such students? "I think the most important thing would be to gradually, without pouncing, try to gain their confidence."

> People love to be flattered. Don't become unrealistic. But flatter them. Compliment them. Because I can assure you that if you're a white male faculty member, you probably are not really aware of the extent to which black students are not complimented. Complimenting encourages. Don't be extravagant to the point that it's ridiculous, but be encouraging. If you can, gain that student's confidence. Try to take a genuine interest; people will see that and they will open up a little bit. This may seem intrusive, but will be appreciated.

ROLE MODELS

It was not unusual for Fern's students to tell her how much they appreciated having a female black instructor. At the time of the interview, she was the only African American woman in the department—despite the fact that Howard University is a predominantly black institution! Fern was written up in a brochure which was used in an exhibit in a Chicago museum. From this she has gotten numerous letters from students who were interested in mathematics and had never seen or heard of a black woman in mathematics; they were hungry for a role model.

Some people object to the idea of a role model because they think that each person is unique, and no one can be a role model for anyone else. But a role model, more than anything else, simply opens doors of possibility. As we see in Fern Hunt's case, for women or minority students who have never seen a mathematician of their sex, race, or ethnic background, it can be crucial to find such a model of possibility.

Education in Black and White

Race has obviously been an important factor in Hunt's education and development. How has this influenced her thoughts about the advantages and disadvantages of predominantly black schools vs. predominantly white institutions?

> *If you had children, would you advise them to go to a predominantly black college?*

Not necessarily. A lot would depend on the personality of the student. There was a point, when I was teaching at Howard, when I would say that they should definitely try to get into a Dartmouth College, for example. At that time there wasn't the racial exclusion that there had been in previous years. The majority of colleges were integrated. They were admitting students of all colors and actively seeking black students. I advised students at that time to try to go to majority colleges, especially with their previous experience of being in predominantly black schools. You need to meet other kinds of people; I felt that was very good and very healthy. We need to do that as a society, we need to get out of our collective ghettos. However, in recent years there has been a growing intolerance, and I have met some of the students who have transferred from majority colleges to Howard. There is a terrific economic anxiety among white students right now, and some of them seem to be taking it out on black students on these campuses.

I also find that sometimes black students are forming little cliques on the majority campuses. So the situation is very complex. I no longer hold the unequivocal position that yes, you should definitely go to the majority, or yes, you should definitely go to a black college; there are pluses and minuses with both choices. It would depend very much on the individual.

In the end, Fern says, there is no clear answer to this difficult question. It depends entirely on the particular student, his or her interest and personality, and the particular institutions in question.

Clearly Fern is just one person, with one set of experiences. But her story illustrates some of the ways that being a woman and being black influence one's mathematical development. Being black affected her experience at Bryn Mawr, which, while all-female, was a predominantly white institution; it influenced the positions that opened to her when she completed her Ph.D.; and it influenced her experiences with her students at Howard, a predominantly black institution.

At the same time, certain lessons emerge that are universal—most notably that support and community are critical factors in achieving success. This is a lesson that has influenced Hunt's style in teaching. It also suggests one direction for improvement in making mathematics a more inclusive endeavor.

6

The Quest for Certain and Eternal Knowledge

Myths: Mathematics is a realm of complete objectivity.
Mathematics is non-human.

The primary service of modern mathematics is that it alone
enables us to understand the vast abstract permanences which
underlie the flux of things.[1]

—George Birkhoff

Two thousand years ago, the Pythagoreans believed that mathematics was
the language of the universe—an idea encapsulated in the simple phrase
"All is number." Even the secret patterns of the sky, the moon, and the sun
could be revealed through the study of number and shape. Hence, to
pursue mathematics was to pursue truth—certain, eternal, and absolute
truth.

In later centuries, mathematics was seen as evidence of God's existence.
Only an intelligent being, it was assumed, could create a universe that
conformed to such beautiful and simple patterns. In this way, mathematics was seen as a noble cause, a way of revealing God's divine will.
Mathematical knowledge was closely tied to knowledge of the divine.

Though mathematics is no longer seen as a religious pursuit, there is
still a reverence for the patterns of nature, and a hunger for knowledge
which has such claims to certainty and immortality, which clearly delin-
eates "right" from "wrong," and which is equated with "truth." And
mathematics continues to satisfy this hunger, for it remains identified with
certain and eternal knowledge. Indeed, this vision of mathematics not

only is held by the uninformed public, it is firmly rooted within the mathematics community as well, perpetuated in how mathematics is taught, written, and promoted.

But what is the basis for this belief that mathematics is certain and eternal knowledge? Is it an accurate description of mathematics? What impact does it have on women? Are there other ways to think about mathematical knowledge? These questions frame the final chapter of this book.[2]

Underlying Assumptions about Mathematics

MATHEMATICS AS CERTAIN AND ETERNAL KNOWLEDGE

What is it that distinguishes mathematical knowledge from other forms of knowledge? Students of all ages, whether they are studying algebra, trigonometry, geometry, or calculus, would agree on one universal theme: Mathematics is beyond doubt. A mathematical statement is either right or wrong, and any true statement can be proven. Moreover, once it is proven true, it can never later turn out to be false. For many people, this kind of knowledge is appealing and comforting; for others it is a source of frustration. But rarely is this vision questioned. Indeed, the belief that mathematics is certain knowledge is precisely what draws many people to mathematics. As Albert Einstein once wrote, "At the age of 12 I experienced a second wonder . . . in a little book dealing with Euclidean plane geometry . . . [there] were [mathematical] assertions . . . which—though by no means evident—could nevertheless be proved with such certainty that any doubt appeared to be out of the question. This lucidity and certainty made an indescribable impression on me."[3]

Few other domains can make this claim; religion is the only real competitor. According to the mathematician J. L. Synge,

> In the common opinion certainty is to be found in two places and only two—religion and mathematics. In religion you believe by faith, in mathematics you prove what you want to prove. Outside these cosy domains, life is full of doubt and uncertainty and that is what people prefer because it is what they are used to. But it is a comfort to feel that, if the doubt and uncertainty become excessive, one can fall back on the eternal verities of religion or mathematics.[4]

However, for those who are not religious, or who have been weaned on the skepticism of Descartes and do not trust any knowledge that cannot be proven through rational thought, mathematics is unparalleled. The mathematician and philosopher Bertrand Russell wrote: "I wanted certainty in the kind of way in which people want religious faith. I thought that certainty is more likely to be found in mathematics than elsewhere."[5]

A second ingredient that distinguishes mathematics is its claims to eternal knowledge. While some have sought immortality through writing, art, offspring, empires, and secret potions, others have sought it in mathematics. As the mathematician G.H. Hardy explains, "What we do may be small, but it has a certain character of permanence; and to have produced anything of the slightest permanent interest, whether it be a copy of verses or a geometrical theorem, is to have done something utterly beyond the powers of the vast majority of men."[6]

Even those who are not seeking their own immortality by proving profound theorems that might be reproduced in texts for centuries to come are drawn to mathematics because it represents eternal truths, a secure and unchanging universe. Bertrand Russell described mathematics as a "refuge in a timeless world, without change or decay or the will-o-the-wisp of progress."[7] And the mathematician Cassius Keyser described mathematics as "the human spirit's craving for invariant reality in a world of tragic change."[8] While people may grow old, decay, and die, mathematics never does; nor does it change moods or faces—mathematics is predictable and reliable.

A third belief about mathematics that is often identified with its claim to certain and eternal knowledge is that it is universal. Mathematics is assumed not only to transcend individuals—whether they be white or black, female or male, young or old—it is also assumed to transcend culture. The Pythagorean Theorem is seen as true whether one is a Pygmy, a Hindu, a Muslim, or an Eskimo. In this vein, the mathematicians Davis and Hersh write that mathematicians regard their work as containing truths which not only are "valid forever" but apply to "the most remote corner of the universe."[9] This universal quality solidifies the belief that mathematics is free of bias. It suggests that mathematical truths cannot be manipulated by culture or politics, and in this way mathematics remains both objective and pure.

The vision of mathematics as universal is often taken yet one step further. Not only is mathematics seen as applying to all people and all cultures, it is assumed to apply to all spheres of inquiry, including philosophy, economics, psychology, and history; it is also applied to business, sports, politics, music, and art. As E. P. Northrop wrote, "Is this not the real place of mathematics in a liberal education—not simply as a subject matter, or as a discipline applicable only to its own subject matter, but as a discipline which is applicable to almost every intellectual activity of man?"[10] In this way, mathematics has a kind of paradigmatic status. It is a model that many other disciplines strive to emulate, for it represents a kind of pure knowledge that other disciplines hunger for.

Given the enormous impact of such a vision of mathematics, it is worth exploring the roots of this vision. How did mathematics come to be seen in this way?

THE FOUNDATION FOR CERTAIN AND
ETERNAL KNOWLEDGE

What are the underlying beliefs that buttress the vision of mathematics as certain and eternal knowledge?

The identification of mathematics with *certain* knowledge has historically rested on two foundations. The first is the belief that mathematics is the language of the universe and so *must* be true. In this context, there is no room for doubt, because mathematics is seen as a direct reflection of our physical reality; and since the facts of the physical world are viewed as certain, so too is mathematical knowledge. Thus, because two apples plus two apples always yields four apples, we assume that 2 + 2 will always equal 4. The real world gives us knowledge of mathematics. Conversely, what is true in mathematics is seen as true in the real world. For example, if we can prove mathematically that the three medians of a triangle intersect in a single point, the assumption is that this must be true in the physical world as well. The mathematical and the physical world are direct mirrors of each other.

The second mechanism for ensuring certainty resides in the *methods* of mathematics: in particular mathematics is equated with formal deductive reasoning. Deductive logic was designed to capture the theoretically universal acknowledgment that certain lines of reasoning are always reliable. For example, any reasonable person would agree that if all women are mortal, and Hypatia is a woman, then there can be no doubt that Hypatia is mortal. Logic is a systematic attempt to categorize these kinds of arguments. Because the assumption is that any rational person would agree with the validity of such arguments, logical proofs are seen as objective and universal. And since the fundamental guide in designing deductive logic was that it preserve truth—i.e., if one's starting assumptions are true, than any logical statement that can be derived from them must also be true—deductive reasoning also preserves certainty. If you start with something certain, you end with something certain.

The vision of mathematics as *eternal* knowledge, on the other hand, rests largely on the dominant philosophy of mathematics called Platonism. Platonism is the belief that there is a mathematical world that in some sense has an independent existence; it is a world of relationships and forms that transcends the human and physical world. As Reuben and Hersh explain:

> According to Platonism, mathematical objects are real. Their existence is an objective fact, quite independent of our knowledge of them. Infinite sets, uncountably infinite sets, infinite-dimensional manifolds, space-filling curves—all the members of the mathematical zoo are definite objects, with definite properties, some known, many unknown. These objects are, of course, not physical or material. They exist outside the space and time of physical existence. They are immutable—they were not created, and they will not change or disappear. Any meaningful question about a mathematical object has a definite answer, whether we are able to determine it or not. According to Platonism, a mathematician is an empirical scientist like a geologist; he cannot invent anything, because it is all there already. All he can do is discover.[11]

Because Platonism defines mathematics as existing independently of the physical world, it preserves its eternal quality: mathematical ideas and relationships do not change or decay. Indeed, this vision implies that any mathematical fact that was true two thousand years ago not only would be true today, it also would be true two thousand years from now.

Platonism is not the only philosophy of mathematics; there are several others.[12] Formalism, for example, which was developed in the early part of this century, views mathematics as a human construction. Its focus is on the symbols and formulas of mathematics rather than mathematical concepts or objects. Formalism takes a syntactic rather than a semantic view. Just as an English sentence can be thought of either as simply a string of letters and words put together in a prescribed way or as a representation of a certain *idea* which is embodied in the meaning of the words, so too can a mathematical formula be seen either as a string of symbols or as a representation of a mathematical concept/object. For example, the sentence "The sun is hot" can be described as consisting of four words put together in a grammatically acceptable way, or it can be thought of as describing some fact of the universe which we know to be true.

> According to formalism . . . there are *no* mathematical objects. Mathematics just consists of axioms, definitions and theorems—in other words, formulas. In an extreme view, there are rules by which one derives one formula from another, but the formulas are not about anything; they are just strings of symbols. Of course the formalist knows that mathematical formulas are sometimes applied to physical problems. When a formula is

given a physical interpretation, it acquires a meaning, and may be true or false. But this truth or falsity has to do with the particular physical interpretation. As a purely mathematical formula, it has no meaning and no truth value.[13]

In Platonism, since mathematics exists independent of human beings, a mathematician's job is to *discover* mathematics. In formalism, a mathematician's job is to *create* mathematics.

The formalist philosophy does not present quite as compelling a picture of mathematics in terms of its claim to absolute and pure knowledge. From a formalist perspective, mathematics becomes almost arbitrary, or at least (like language) the product of convention. There is no single, incontrovertible reality. The effectiveness of mathematics in the real world becomes an almost accidental coincidence. In such a philosophy it also becomes much harder to decide which set of rules or axioms to follow. And the criteria used to determine what constitutes important mathematics must shift; in fact, it is surprisingly difficult to define what criteria to use.

While many mathematicians avoid being pinned down to a single philosophical belief system, in particular whether mathematics is discovered (Platonism) or invented (formalism), their practice most often is aligned with a Platonist perspective typified by the words of G. H. Hardy: "I believe that mathematical reality lies outside us, that our function is to discover or *observe* it, and that the theorems which we prove, and which we describe grandiloquently as our 'creations,' are simply our notes of our observations."[14]

However, while most mathematicians operate on this basic assumption, some are more guarded in publicly committing to this view, as Reuben and Hersh put it: "The typical mathematician is a Platonist on weekdays and a formalist on Sundays. That is, when he is doing mathematics he is convinced that he is dealing with an objective reality whose properties he is attempting to determine. But then, when challenged to give a philosophical account of this reality, he finds it easiest to pretend that he does not believe in it after all."[15]

On a pragmatic level, one reason the Platonist philosophy is held so tenaciously is that it fuels mathematics' claim to complete objectivity. According to Webster's Dictionary, "objective" is defined as

- existing as an object or fact, independent of the mind; real
- concerned with the realities of the thing dealt with rather than the thoughts of the artist, writer, etc.
- without bias or prejudice

Objectivity in mathematics, then, implies that mathematics has an independent existence, that the focus of mathematics is on mathematical "objects" rather than the people who do mathematics, and that mathematics is free of bias or prejudice. All of these qualities are built into Platonism.

The Platonist philosophy serves another purpose as well. Because it feeds the image that mathematics is removed from the human and physical world, mathematics can in some sense remain pure. Indeed, many mathematicians see mathematics and the human world not only as separate, but also as hierarchically ordered: mathematics represents what is pure and beautiful, whereas the human world represents something far inferior. As G. H. Hardy writes, the "very remoteness [of mathematics] from ordinary human activities should keep it gentle and clean."[16] For Bertrand Russell, the separation of the two spheres was a central aspect of what drew him to mathematics:

> Mathematics, rightly viewed, possesses not only truth, but supreme beauty—a beauty cold and austere, like that of sculpture, without appeal to any part of our weaker nature, without the gorgeous trappings of painting or music, yet sublimely pure, and capable of a stern perfection such as only the greatest art can show. . . .
>
> Real life is, to most men, a long second-best, a perpetual compromise between the ideal and the possible; but the world of pure reason knows no compromise, no practical limitations, no barrier to the creative activity embodying in splendid edifices the passionate aspiration after the perfect from which all great work springs. Remote from human passions, remote even from the pitiful facts of nature, the generations have gradually created an ordered cosmos, where pure thought can dwell as in its natural home, and where one, at least, of our nobler impulses can escape from the dreary exile of the actual world.[17]

David Bloor in his book *Knowledge and Social Imagery* argues that the separation of the two worlds is often seen as necessary in order to prevent the pure world of mathematics from becoming polluted by the everyday world. In the context of science more generally, he writes: "Science is sacred, so it must be kept apart. It is, as I shall sometimes say, 'reified' or 'mystified.' This protects it from pollution which would destroy its efficacy, authority and strength as a source of knowledge."[18] The same reasoning applies to mathematics. As Bertrand Russell writes, mathematics is "an abstract edifice subsisting in a Platonic heaven and only reaching the world of sense in an impure and degraded form."

Thus, there are three central foundations to the belief that mathematics is certain and eternal knowledge: the assumption that mathematics is in some sense a direct reflection of our physical world, the reliance on formal deductive reasoning, and the Platonic vision of mathematics as an independently existing reality. But are these foundations sustainable?

Challenges to Underlying Assumptions

PROBLEMS WITH THE BELIEF THAT MATHEMATICS = PHYSICAL WORLD

For thousands of years it was believed that Euclidean geometry was *the* geometry of the universe. And because it was assumed that there was only *one* geometry, the very axioms or postulates that this geometry was based on were seen as self-evident truths. They were chosen to represent the most elemental basis of geometry, whose truth could not be doubted. From this absolute and certain foundation, all else could be deduced, with the same claim to certainty that these fundamental starting postulates had. This deep and fundamental belief that there was one geometry, and that it was the geometry of the universe, remained unquestioned for centuries.

But the introduction of non-Euclidean geometry in the nineteenth century had a profound effect on the world of mathematics. The recognition that there could be more than one geometry, and that each could be as consistent and even as useful in real-world applications as the others, led to a completely different conception of mathematics and its relationship to the universe. It was not just a matter of subsuming Euclidean geometry into the new enlarged geometry, for the "facts" of Euclidean geometry could be in direct contradiction with the facts of other non-Euclidean geometries. For example, in Euclidean geometry, the angles of a triangle add up to a sum of 180 degrees; in non-Euclidean geometry, they can add up to more or less than 180 degrees. Moreover, even if Euclidean geometry is seen as a special case of our more generalized vision of geometry, what had irrevocably changed is the belief that there is one, and only one, geometry of the universe. Accompanying this radical shift was the recognition that one could no longer take a set of postulates as self-evident, certain truths. Suddenly it was possible to work with many different sets of starting axioms, each leading to different kinds of geometry, different theorems, different truths.

Thus, the idea that Euclidean geometry was essentially a mathematical map of the universe had proven to be false. Indeed, what had been sacrificed was also mathematics' claim both to certain and to universal knowledge. With this dramatic shift came the seed of a very different vision of mathematical knowledge, a vision that sees it as contextual rather than universal. In this new vision, it makes sense to talk about mathematical truths only within a certain context of starting assumptions. The implications of this shift are even now not fully incorporated into the practice of mathematics.

In a similar way, the discovery, or invention, of new algebraic systems (in particular the development of the quaternions) challenged the belief in the certainty of algebraic rules, which had been considered as fundamental as the rules of Euclidean geometry. Because basic assumptions such as that $axb = bxa$ did not necessarily hold in other mathematical frameworks, it was no longer clear that *any* statement could be made with complete certitude. Thus, neither geometry nor algebra was safe from this revision of mathematics as relativistic rather than universal and certain knowledge.[19]

PROBLEMS WITH THE BELIEF THAT
MATHEMATICS = DEDUCTIVE REASONING AND
FORMAL PROOFS

Even if it is true that mathematics is not a direct reflection of the physical world—or at least that we can no longer think in terms of there being only one mathematical universe, which mirrors the physical universe—it would seem that at least the *methods* of mathematics are still impeccably reliable. Since mathematics is based on formal deductive reasoning, which ensures that truths follow from truths, something that was once proven to follow from a set of assumptions cannot later be proven to be false in that context. In this way, mathematics can still maintain claims to a certain kind of incontrovertible knowledge—knowledge that is objective, reliable, and certain. However, there are four possible objections to the belief that mathematics can be equated with formal deductive reasoning.

First Objection: Mathematicians Don't Really Work with
Formal Deductive Proofs

While in theory any mathematical proof can be described as a formal deductive proof, in reality it almost never is. In practice, proofs are simply whatever it takes to convince colleagues that a mathematical idea is true. In this vein, the mathematicians Davis and Hersh make the distinction between formal and informal mathematics. Formal mathematics involves formal proofs: a string of statements each of which is an axiom or follows from previous statements by specified rules of logic. And indeed, these kinds of formal proofs have the closest claims to certainty. However, as

Davis and Hersh point out, formalized mathematics "is in fact hardly to be found anywhere on earth or in heaven outside the texts and journals of symbolic logic."[20] In fact, as Russell and Whitehead showed in *Principia Mathematica*, it takes nearly a hundred pages to prove that $1 + 1 = 2$ in this rigorous a fashion.[21]

Informal mathematics, on the other hand, is the way mathematics is normally practiced. It involves more informal arguments which are used to persuade others that an idea or result is correct. But this kind of mathematics *does* change and evolve; it can even be false. As Davis and Hersh put it, "Mathematics, too, like the natural sciences is fallible, not indubitable; it too grows by the criticism and correction of theories which are never entirely free of ambiguity or the possibility of oversight."[22]

Thus, the certainty of mathematics is tied to an idealized version of mathematics that is far from the reality practiced. Proofs are rarely the rigorous deductions defined in logic classes; they are something far looser and more fallible.

Second Objection: What Counts as a Proof?

Moreover, what constitutes an acceptable proof in the community changes over time. As Israel Kleiner, in his article "Rigor and Proof in Mathematics: A Historical Perspective," points out, "The notion of proof is not absolute. Mathematicians' view of what constitutes an acceptable proof have evolved." Indeed, he argues that "the validity of a proof is a reflection of the overall mathematical climate at any given time." For ancient Greeks, rigor was of the utmost importance. Mathematical proof became equated with precise mathematical deduction. But centuries later, standards of rigor had changed dramatically. Many of the great developments in calculus took place in the midst of, and perhaps because of, very loose standards of rigor.[23]

It is not only the standards of rigor that have changed in the past two thousand years, but also the tools used to establish rigor.

> In ancient Greece, a theorem was not properly established until it was geometrized. In the Middle Ages and the Renaissance, geometry continued to be the final arbiter of mathematical rigor (even in algebra). Mathematicians' intuition of space appeared, presumably, more trustworthy than their insight into number—a continuing legacy of the consequences of the "crisis of incommensurability" in ancient Greece. The calculus of the 17th and especially the 18th century was no longer easily justifiable in geometric terms, and algebra became the major tool of justification (such as there was). There was a mix of the algebraic and geometric in Cauchy's work. With Weierstrass and Dedekind in the latter part of the 19th century, arithmetic rather than geometry or algebra had become the language of

rigorous mathematics. To Plato, God ever geometrized, while to Jacobi, He ever arithmetized. The logical supremacy of arithmetic, however, was not lasting. In the 1880s Dedekind and Frege undertook a reconstruction of arithmetic based on ideas from set theory and logic.[24]

Even today, the question of what constitutes a proof is a hotly debated topic. As John Horgan argues in his intriguing article "The Death of Proof," two important factors are challenging traditional notions of proof.[25] One is the extreme complexity of many modern proofs. Andrew Wile's proof of Fermat's Last Theorem, for example, is two hundred pages long, and only a tiny percentage of mathematicians would be qualified to evaluate it. As Horgan writes, "Wiles' claim was accepted largely on the basis of his reputation and the reputations of those whose work he built on. Mathematicians who had not yet examined the argument in detail nonetheless commented that it 'looks beautiful' and 'has the ring of truth.'"[26] Another important result, the classification of finite simple groups, was completed in the early 1980s. It took five hundred articles totaling nearly fifteen thousand pages and more than one hundred workers to prove it. Clearly this level of complexity makes it hard for anyone to be sure whether the proofs are correct. As Horgan says of the classification theorem, "It has been said that the only person who grasped the entire proof was the general contractor, Daniel Gorenstein of Rutgers."[27] But since Gorenstein died last year, one is left wondering if there is *anyone* who can be sure that the proof is correct. Given these examples, it is not surprising to learn that in many instances there is intense disagreement, even among eminent mathematicians, about whether a proof is correct. The idea that the truth or falsity of a proof can be definitively and objectively determined is based on a tradition of proofs' being relatively short and easy to verify by other mathematicians. When proofs are this long and complex, such verification is almost impossible. Is it appropriate, then, to say that a result has been objectively established to be true?

A second factor that is dramatically influencing current visions of proof is the computer. More and more mathematicians rely heavily on computers as both a tool to generate ideas and an aid in proving theorems. The proofs of some theorems involve checking hundreds or thousands of cases of some mathematical configuration. In these cases human calculations or confirmation is unrealistic. For these reasons, mathematics is becoming much closer to an experimental science, and as with all experimental sciences, claims to certainty cannot be ensured. Human, machine, or design errors can enter in at any stage.

Furthermore, computers are also being used to check the accuracy of proofs; in some cases this yields a *probability* of a proof's being correct. In this way, our notions of both mathematical proof and truth are being transformed. Rather than being able to say that a proof is absolutely either true or false, we may increasingly hear phrases such as "This theorem is true with probability *p*." This is a far cry from the vision of absolute certainty inherited from the Greeks.

Finally, even if proofs *were* translated into formal deductive proofs, there is not universal agreement about which rules of logic to use. Different mathematical schools of thought have different ideas about which laws of logic are acceptable. The adherents of one school of thought, represented by the constructivists (or intuitionists), "regard as genuine mathematics only what can be obtained by a finite construction."[28] For example, they do not accept the set of natural numbers (counting numbers) as a valid mathematical object because it contains an infinite number of members. Furthermore, they reject the law of the excluded middle (that if *p* is a mathematical statement, then either *p* or not *p* must be true), and they reject proofs by contradiction. However, other schools of thought, such as the formalists, strongly disagree. The differences between these schools are significant. Large segments of mathematical research would be rejected by one school, but accepted by another. As Israel Kleiner argues,

> The differences between the formalists and the intuitionists were genuine. For the first time mathematicians were seriously (and irreconcilably) divided over what constitutes a proof in mathematics. Moreover, this division seems to have had an impact on the work that at least some mathematicians chose to pursue, as the testimony of two of the most prominent practitioners of that epoch—J. von Neumann and H. Weyl, respectively—indicate:[29]
>
>> In my own experience . . . there were very serious substantive discussions as to what the fundamental principles of mathematics are; as to whether a large chapter of mathematics is really logically binding or not. . . . It was not at all clear exactly what one means by absolute rigor, and specifically, whether one should limit oneself to use only those parts of mathematics which nobody questioned. Thus, remarkably enough, in a large fraction of mathematics there actually existed differences of opinion!
>>
>> Outwardly it does not seem to hamper our daily work, and yet I for one confess that it has had a considerable practical influence on my mathematical life. It directed my interests to fields I considered relatively "safe," and has been a constant drain on the enthusiasm and determination with which I pursued my research work.

Thus, the idea that mathematical knowledge is certain and completely objective relies on the belief in a clear, universal notion of mathematical proof. But as we see, the idea of proof is evolving, controversial, and subject to social negotiation. As the Field's Medalist William Thurston says, "That mathematics reduces in principle to formal proofs is a shaky idea, [peculiar to this century]. In practice, mathematicians prove theorems in a social context. It is a socially conditioned body of knowledge and techniques."[30]

Third Objection: Formal Deductive Systems Are Inherently Limited

Yet another important challenge to the identification of mathematics with formal deductive reasoning emerges from the monumental work of Kurt Gödel in the first half of this century. Gödel showed that within any kind of formal system at least as powerful as arithmetic, there were mathematical statements which we knew to be true but which could not be proven within that system. So just as we were adjusting to the idea that mathematical ideas are contextual, along came the shattering news that these contexts were necessarily, and inherently, limited in their ability to formally prove all that was true of them. Once again, a long-held belief— one which was assumed to be an essential and eternal aspect of mathematics, that all mathematical truths could be formally proved—was being shaken.

As if this were not enough, Gödel also established that one could not formally prove within such a formal system that the system itself was consistent, i.e., free of contradictions. In mathematics, contradictions are a death knell, like an infectious black plague that cannot be contained. For once *one* contradiction exists, it can be proven mathematically that *any* mathematical statement can be proven true in that system; i.e., anything goes.[31] Since even one contradiction obliterates any hope of having certain knowledge, mathematicians wanted desperately to prove the consistency of mathematics. And many of the most prominent mathematicians around the turn of the century devoted themselves to this cause. However, Gödel's work was a shattering blow to that effort. As Morris Kline summarizes, "It is now apparent that the concept of a universally accepted, infallible body of reasoning—the majestic mathematics of 1800 and the pride of man—is a grand illusion. Uncertainty and doubt concerning the future of mathematics have replaced the certainties and complacency of the past."[32]

Fourth Objection: The Focus on Deductive Reasoning
Gives an Inaccurate Picture of Math;
Intuition Plays Just As Important a Role As Formal Proofs

There is one final objection to the identification of mathematics with formal deductive reasoning—namely that it presents an artificially narrow view of what mathematics is all about, and in this way is misleading, if not inaccurate. Mathematical proofs are the mere byproducts of a complex and subtle process that involves not only informal reasoning but a great deal of intuition and insight. As the mathematician Paul Halmos explains,

> Mathematics . . . is never deductive in its creation. The mathematician at work makes vague guesses, visualizes broad generalizations, and jumps to unwarranted conclusions. He arranges and rearranges his ideas, and he becomes convinced of their truth long before he can write down a logical proof. The conviction is not likely to come early—it usually comes after many attempts, many failures, many discouragements, many false starts. It often happens that months of work result in the proof that the method of attack they were based on cannot possibly work, and the process of guessing, visualizing, and conclusion-jumping begins again. A reformulation is needed . . . more experimental work is needed. . . . The deductive stage, writing the result down, and writing down its rigorous proof are relatively trivial once the real insight arrives; it is more like the draftsman's work, not the architect's.[33]

The link between reason and intuition is even more explicitly described by Morris Kline:

> Mathematicians create by acts of insight and intuition. Logic then sanctions the conquests of intuition. It is the hygiene that mathematics practices to keep its ideas healthy and strong. Moreover, the whole structure rests fundamentally on uncertain ground, the intuitions of man. Here and there an intuition is scooped out and replaced by a firmly built pillar of thought; however, this pillar is based on some deeper, perhaps less clearly defined intuition. Though the process of replacing intuitions by precise thoughts does not change the nature of the ground on which mathematics ultimately rests, it does add strength and height to the structure.[34]

However, in Western society, reason is characteristically pitted against intuition.[35] The two are seen as contrasting, even conflicting, ways of understanding the world. Moreover, mathematics is traditionally identified exclusively with formal reasoning. Why? Part of the reason can be found in tracing the history of mathematics and the problems that intuition introduced. Intuition, at times, led to contradictions, paradoxes, or false conclusions. Thus, in a desire to free mathematics from these perils, some sought to ban intuition altogether. As the mathematician Hans Hahn explains,

> Because intuition turned out to be deceptive in so many instances, and because propositions that had been accounted true by intuition were repeatedly proved false by logic, mathematicians became more and more skeptical of the validity of intuition. They learned that it is unsafe to accept any mathematical proposition, much less to base any mathematical disci-

pline on intuitive convictions. Thus a demand arose for the expulsion of intuition from mathematics reasoning, and for the complete formalization of mathematics. That is to say, every new mathematical concept was to be introduced through a purely logical definition; every mathematical proof was to be carried through by strictly logical means. . . .

The task of completely formalizing mathematics, of reducing it entirely to logic, was arduous and difficult; it meant nothing less than a reform in root and branch. Propositions that had formerly been accepted as intuitively evident had to be painstakingly proved.[36]

As history suggests, however, an overreliance on reason can be just as problematic as an overreliance on intuition. Indeed, intuition has always been, and will always be, a necessary element of mathematical progress—even, at times, at the expense of rigor. As mentioned earlier, a classic example of this is found in the history of calculus, whose development might have been seriously hampered by Hahn's extreme skepticism of intuition and non-rigorous mathematics. As David Bloor writes about this period in which calculus was being created,

Historians looking back on this fruitful period sometimes remark on the lack of rigour attending the use of infinitesimals. . . . On the other hand historians have certainly acknowledged the value of the decline in rigour which allowed these terms for the first time to feature explicitly in calculations. Before this time they were forbidden, and they are forbidden now. As the historian [Carl] Boyer says, "luckily" men like Wallis did not worry too much about rigour.[37]

In this case, loosening of rigor was essential for creativity.

Similarly, the famous and prolific Indian mathematician Ramanujan did not abide by traditional standards of mathematical rigor. In his case it was because of his limited access to formal training in mathematics, but his example illustrates not only how important intuition is to mathematical work, but also that the line between reason and intuition can be unclear.

Since Ramanujan's college education spanned but one year, he did not have a firm grasp of what constitutes a rigorous proof. Ramanujan's notebooks and unpublished papers, as I have noted, provide only a few scattered proofs. Even his published papers contain several unverified statements. In certain instances, mathematicians can ascertain his arguments, and frequently they are not rigorous. Ramanujan would often employ various techniques and procedures that needed to be justified. In other words, some applications are valid, and others are not; but, despite the shortcomings of his proofs, Ramanujan very seldom committed serious errors. He had an uncanny ability to determine when his methods produced correct results and when they did not.[38]

Given the critical role that both intuition and reason play, why are they not equally emphasized in mathematics education and in the public images of mathematics? One reason that the mathematics community is not more active in conveying a more accurate picture of mathematics is that much of the power and prestige of mathematics comes from its claim to certainty and its image as an "exact science." Intuition, on the other hand, seems vague and fuzzy; it is therefore relegated to the private world of mathematics, or (in some attempts) exorcised completely.

Many other disciplines in the past few decades have sought to model themselves after mathematics. Fields such as history, philosophy, and sociology have attempted a more rigorous, formal, and quantitative approach to research—all because mathematical knowledge is seen as the closest to pure truth, and the method of logical reasoning is seen as the vehicle to attain such truth. But would mathematics maintain quite this level of extreme admiration and emulation if others had a more complete and realistic picture of it? As the mathematician Carl Allendoerfer wrote, "The inner circle of creative mathematicians have a well-kept trade secret that in a great many cases theorems come first and axioms second. This process of justifying a belief by finding premises from which it can be deduced is shockingly similar to much reasoning in our daily lives, and it is somewhat embarrassing to me to realize that mathematicians are experts at this art."[39]

Although many mathematicians argue that formal proofs are a kind of afterthought to the process of doing mathematics, it is this identification with formal proofs that gives mathematics its reputation and prestige. Hence formal reasoning continues to play a dominant role in the public imagery of mathematics. Indeed, the image of mathematics as certain knowledge has such a powerful hold that any threat to this vision of certainty means nothing less than "an expulsion of intuition from mathematics reasoning." Since intuition can lead mathematicians into dangerous, even false, waters, and since it is reason that keeps intuition in line, it is reason, rather than intuition, that has become the public hallmark of mathematics. And it is reason that is emphasized in mathematics education.

However, the problem with equating mathematics with pure deductive reasoning is that it is simply inaccurate and misleading. Formal proofs are simply the last stage of a far more unpredictable, uncertain, and non-formalizable process. The knowledge gained from this process is not certain and infallible, but rather tentative, approximate, and evolving.

PROBLEMS WITH PLATONISM

One challenge to the Platonist belief that there is a mathematical world that has an independent existence has already been presented. The introduction of non-Euclidean geometry undermined the belief that there is a single mathematical universe; suddenly there were many different compelling geometric systems that worked, and yet that contradicted each other. Which, then, is supposed to be an accurate description of the Platonic realm? And if they are all true, how could contradictory things hold in this mathematical realm? If one contradiction is permitted, then infinitely many more follow, and the whole mathematical system is undermined. It is unclear, therefore, how to reconcile a Platonic view with these conflicting mathematical systems.

The second problem with Platonism is of a different nature, namely that it is misleading. It fuels the idea that mathematics is completely objective—apolitical and value-free. It removes any traces of human fingerprints. Indeed, for many mathematicians this is an appealing feature of Platonism. By positing the existence of a mathematical universe that transcends both individuals and culture, mathematics assumes a universal quality that can also transcend human bias. As Bertrand Russell wrote at an early stage of his career, "Mathematics takes us still further from what is human, into the region of absolute necessity, to which not only the actual world, but every possible world must conform."[40]

Ultimately, it is misleading to speak of "pure mathematics" untainted by human values, bias, and customs, and it is impossible to separate the product from the process, or mathematics from the mathematician or the mathematical community. Like all intellectual activity, mathematics has subjectivity woven into the fabric of its existence.

The vision of mathematics as purely objective emerges when the focus is simply on the product of mathematical activity: namely the final theorems and proofs. However, a more realistic picture emerges when we see these in a larger context. The theorems can be likened to the flowers of a plant. To see the flower (or theorem) as an independently existing phenomenon of nature, as compelling and beautiful as they may be, deprives us of understanding the complexity of the system that gives that flower its life and sustains it. There is the leaf on which the flower blossoms (namely the proof), the branch that sustains the leaf (the mathematician who got the result), the main stem (the larger mathematics community, funding sources, etc.), the roots of the plant (the large body of mathematics on which this result is based), and the tiny microscopic hairs and nodes which feed on the surrounding soil (society at large). Eventually the seeds of the flower may fall to the ground, giving life to new plants and new flowers. However, the seed will give rise to new fruit only if the soil is fertile.

Many mathematicians might grant that mathematics does indeed need fertile soil to blossom. They could cite centuries of what appears to be mathematical stagnation, and other periods of incredible productivity. They would probably also acknowledge that social conditions are important in stimulating these prolific periods. For example, the industrial revolution coupled with the Renaissance fueled the development of the powerful tools and ideas of calculus. These mathematicians recognize and applaud our current system of universities, their focus on research, and the government support of research through NSF, DOD, and other national agencies. But they would also argue that mathematics is like a seed waiting to blossom; it's all in there, and as soon as the conditions are reasonable, it will take off. The soil does not change the seed, it merely allows it to realize its potential, to display what was inside, latent, waiting to flower.

What this argument masks, however, is that it is not just the flower and the soil that are involved. What so often becomes invisible is everything in between. The plant is the critical mediator between the soil and the flower. It is the plant that derives the necessary nutrients from the soil, converting them to a usable form—just as the mathematics community takes resources of the society and converts them into sustainable nutrients for mathematicians. Similarly, mathematicians feed on mathematical activity—they sift through ideas and distill them into insights which flower into theorems. What are the ways, then, that culture can shape mathematics?

How Do We Define Important Mathematics?

The mathematician and philosopher Gian Carlo Rota wrote, "Whereas the facts of mathematics, once discovered, will never change, the method by which these facts are verified has changed many times in the past, and it would be foolhardy to expect that it will not change again at some future date."[41] However, culture plays a powerful role in shaping not only the methods used, but also the questions that get asked, and the mathematical avenues pursued. For example, in societies where commerce was central, mathematics historically developed in the direction of number systems and arithmetic. In cultures where astronomy or navigation was central, trigonometry was developed. Islamic mathematics was focused on algebra in order to deal with tax assessment and dividing inheritance as prescribed by Koranic law.[42] Calculus emerged to address the societal concerns that involved quantities that changed continuously over time (to describe, for example, the motion of planets, projectiles, or light). Similarly today, a large portion of research is guided by the needs of the primary funders, and to a large extent this is tied to military and defense needs. Clearly, the values of a culture can have a significant impact on the direction that mathematical research will take. Not all cultures have the same values; hence not all cultures will pursue the same mathematical paths. While particular mathematical results may be universal, what counts as important and the directions of mathematical inquiry are not universal.[43]

How then, do we decide what is important mathematics?

Decisions about what problems to work on are made at many levels: on the individual level, where personal taste and ability are critical factors; in one's immediate mathematics community, where there is discussion about "hot" problems and what other people in one's immediate field are working on; at the wider level of public recognition: what topics are seen as worthy of publication, awards, presentations; and at the largest level: what areas and topics are funded by agencies such as the National Science Foundation.

At each of these levels, decisions are guided by people's preferences, by their sense of what is important or exciting, or by societal priorities, such as applications to military, medical, or industrial needs. But there is no absolute, objective criterion for what defines "important mathematics."

Not only is it extremely difficult to come up with criteria for determining importance, but even if one could, these criteria could not be implemented in a standard way. Determining what is "important" is where peer review plays such a critical role, and where subjective and social forces are strongly and inevitably at play in shaping mathematics.[44]

As an experiment in a recent class I taught, students were asked to imagine that they were directors of NSF and had a budget of several billion dollars. First they made individual decisions about how they would allocate money, and defined projects that they thought were important. Then they formed groups of five, and had to negotiate the allocation of funds as a group and come up with a list of funding priorities. It was a fascinating experiment, first to see the projects they came up with as individuals. The gender divisions were striking. The males consistently put a very high priority on space exploration and high-technology projects. The women predominantly focused on health issues and research into genetic engineering. Working in groups also gave them a clear sense of how subjective these decisions are, and the complexity of this kind of social negotiation.

Even the most renowned mathematicians have difficulty articulating what criteria to use in defining "important" mathematics. G. H. Hardy argues that a mathematician is a maker of patterns, and that beauty and seriousness are the criteria by which his patterns should be judged. But what does Hardy mean by "seriousness" and "beauty"?

> A "serious" theorem is a theorem which contains "significant" ideas. . . . We can recognize a "significant" idea when we see it, . . . but this power of recognition requires a rather high degree of mathematical sophistication, and of that familiarity with mathematical ideas which comes only from many years spent in their company. . . . There are two things at any rate which seem essential, a certain *generality* and a certain *depth;* but neither quality is easy to define at all precisely.[45]

Hardy has replaced one elusive concept, "seriousness," with two elusive concepts, "generality" and "depth." But he attempts to analyze these as well. He argues that generality should include the fact that the theorem should be useful in many other proofs, should connect many other mathematical ideas, and should be generalizable to many other theorems. Perhaps we are finally getting somewhere. But Hardy is quick to hedge even on these criteria: "Some measure of generality must be present in any high-class theorem, but *too much* tends inevitably to insipidity. . . . A property common to too many objects can hardly be very exciting, and mathematical ideas also become dim unless they have plenty of individuality."[46]

Hardy finds it even more difficult to precisely capture the second criterion for determining "serious" mathematics—the depth of a theorem.

> It has *something* to do with *difficulty;* the "deeper" ideas are usually the harder to grasp: but it is not at all the same. . . .
>
> It seems that mathematical ideas are arranged somehow in strata, the ideas in each stratum being linked by a complex of relations both among themselves and with those above and below. The lower the stratum, the deeper (and in general the more difficult) the idea. Thus the idea of an "irrational" is deeper than that of an integer; and Pythagoras's theorem [irrationality of the square root of 2] is, for that reason, deeper than Euclid's [that there are infinitely many primes].[47]

Thus significance, like beauty, is in the eye of the beholder. He goes on to give examples of different levels of depth in mathematical theorems, but in the end says, "I could multiply examples, but this notion of 'depth' is an elusive one even for a mathematician who can recognize it, and I can hardly suppose that I could say anything more about it here which would be of much help to other readers."[48]

Finally, Hardy turns to the most elusive criterion of all in identifying important mathematics, namely the criterion of "beauty." He writes, "A mathematician's patterns, like the painter or the poet's, must be *beautiful;* the ideas, like the colours or the words, must fit together in a harmonious way. Beauty is the first test: there is no permanent place in the world for ugly mathematics."[49] In a similar vein, Poincaré asserted that "the aesthetic rather than the logical is the dominant element in mathematical creativity."[50]

But mathematical beauty is a stubbornly elusive concept. As Davis and Hersh explain,

> Blindness to the aesthetic element in mathematics is widespread and can account for a feeling that mathematics is dry as dust, as exciting as a telephone book. . . . Contrariwise, appreciation of this element makes the subject live in a wonderful manner and burn as no other creation of the human mind seems to do. . . .
>
> But the notion of the underlying aesthetic quality remains elusive. Aesthetic judgments tend to be personal, they tend to vary with cultures and with the generations. . . .
>
> Aesthetic judgment exists in mathematics, is of importance, can be cultivated, can be passed from generation to generation, from teacher to student, from author to reader. But there is very little formal description of what it is and how it operates.[51]

Even when mathematicians acknowledge the elusiveness of defining criteria for determining important mathematics, they try to take refuge in the belief that these judgments, though difficult to make precise, are universal—at least by "informed" judges. As Halmos writes, "The criterion for quality [in mathematics] is beauty, intricacy, neatness, elegance, satisfaction, appropriateness—all subjective, but all somehow mysteriously shared by all."[52]

One is reminded here of the psychology experiment in which a group of six people are shown a series of lines that start out horizontal and become more and more vertical. Unbeknownst to the one genuine subject, the other five members of the group are all cooperating with the experimenter. They are told to keep insisting that the lines are horizontal, in order to see how long the subject will continue to agree with them, contrary to his or her own visual evidence. The experimenters were shocked to discover that the line had to be almost completely vertical before the subject would disagree with the others in the group. Furthermore, even when the experiment was explained afterward, the subject would insist that he or she had indeed seen horizontal lines. This gives us a sense of the incredible influence of a group mind on an individual's judgment, vision, and interpretation of reality. Similarly, in mathematics what is assumed to be an objective evaluation of what counts as important is, in fact, inevitably influenced by individual preference and social norms.

Clearly, then, there are problems with all three of the foundations of the dominant beliefs about mathematics. Thus we are left with the following puzzle: Despite the fact that most of these challenges are well more than fifty to a hundred years old, the vision of mathematics as certain, eternal, and absolute knowledge remains pervasive today. Why do these beliefs continue? One major reason seems to be that mathematics continues to be extremely effective. Some have coined the phrase "the unreasonable effectiveness of mathematics."[53] Though mathematicians do not understand the interface between their abstract ideas, which seem to have a life of their own, and the physical world, they are content to pursue it for its intrinsic and compelling nature. In addition, there remains a persistent optimism among some mathematicians that future developments in mathematics will explain and account for the troubles that have arisen in the past century in the foundations of mathematics. Perhaps. But others have begun to imagine a very different view of mathematics. As the mathematician Hermann Weyl, who has been described as one of the greatest mathematicians of this century, writes, "The question of the foundations and the ultimate meaning of mathematics remains open; we do not know in what direction it will find its final solution or even whether a final objective answer can be expected at all. 'Mathematizing' may well be a creative activity of man, like language or music, of primary originality, whose historical decisions defy complete objective rationalization."[54]

The Impact on Women of the Ideology of Mathematics as Certain and Eternal Knowledge

What do these underlying assumptions about mathematics, the foundations for these assumptions, and a variety of challenges to these foundations have to do with women? There are two ways in which to consider the impact of these beliefs. First, visions of mathematics become a filter that influences who would choose to pursue mathematics and whom the mathematics community takes in as one of its own. For a variety of reasons, women are more likely to be kept out by such a filter. Second, current visions of mathematics can also affect the lives of women already in mathematics, albeit in a more indirect way.

Visions of Mathematics as a Filter

Both the Platonist philosophy and the identification of mathematics with formal deductive proofs fuel an image of mathematics as non-human and impersonal. Because the final product, rather than the process of doing mathematics, is what is highlighted, people are seen as almost irrelevant. Proofs are supposed to weed out the personal—they can be replicated by anyone, and are full proof knowledge. This is the only exposure most people have to mathematics. It is an image perpetuated to a large degree by the way mathematics has traditionally been taught in secondary schools.[55] However, as Douglas Campbell describes in his essay "The Curse of Euclid," it gives a warped vision of mathematics:

> It is a great disservice to students to emphasize how the Greeks used deductive reasoning as a procedure to obtain truth. Euclid's Elements perpetuates the image of mathematicians systematically writing down a proof of some Truth directly from axioms and previous theorems. The proofs of Euclidean geometry are held up as models for reasoning. Students are given a theorem such as "the medians of a triangle meet at a point," and a proof consisting of a series of assertions, each one backed up by some axiom, with the last assertion the desired conclusion. The student begs the teacher to explain why the proof proceeds as it does or to explain how the proof was created—all usually to no avail. Euclid becomes a curse to reasoning, not a blessing. Geometry is usually presented as a series of examples of what constitutes correct reasoning but rarely is it presented in a manner which will teach people how to reason.
>
> No mathematician creates a proof by staring at axioms. He draws pictures, measures, experiments, looks at extreme cases, finds analogies, asks authorities, and makes three brilliant but false starts before he finally gets a flash that sets him up for a proof that works. Then he refines the proof, erases unessential steps, and removes the false starts that led him to the true insight. Finally, he has a polished jewel, a tightly-reasoned, correct, but virtually opaque proof ready to frustrate both student and teacher.
>
> Euclid's axiomatic procedure is a breakthrough; it is a procedure for the unification of material. It allows key assumptions to stand out. It allows for systematic procedures of verification. But so long as students are misled into believing that the polished jewels are the actual reasoning rather than the end product of reasoning, just so long will it be that Euclidean geometry will remain a curse rather than a blessing to the teaching of reasoning.[56]

Not only are students left with the impression that mathematics is a cold, impersonal, polished jewel, they do not develop the tools that would enable them to envision themselves as the creators of that jewel. The identification of mathematics exclusively with deductive reasoning feeds a preoccupation with right and wrong, true and false. If students are taught that all that really matters is getting the right answer, they lose the sense of play that is so important to real learning. The false starts, mistakes, and guesses are as much a part of doing mathematics as getting the right answer. They are a way of exploring the mathematical terrain around a specific problem. Often digressions into what seems like the wrong path can lead to new and important mathematics.

The vision of mathematics as divorced from humans also fuels an approach to teaching mathematics that can be alienating to many students. Traditionally mathematics is presented in a way that strips it of all context: either the *mathematical* context—how and why an idea developed, its significance and relation to other ideas of the time; or its *cultural* context—why a particular avenue of research flourished, and what needs or values it served in the larger culture. Since the Platonic philosophy implies that individuals or cultures are, for the most part, irrelevant to mathematical knowledge, the particular mathematical or cultural context that gave rise to the ideas is seen as peripheral—excess baggage that can be stripped away so as not to distract one's focus from the essence of mathematics: the final theorems and proofs.

When mathematics becomes dehumanized and decontextualized, it is more likely to be seen as irrelevant to students' lives, further discouraging them from pursuing mathematics. They see little connection between mathematical ideas and people's lives—even the people who created the ideas—hence it is hard for students to make connections to their own lives. Mathematics is often viewed as an abstract body of knowledge to be memorized and applied in a mechanistic way. What is missing is the beating heart that pumps life into the material: Why were these ideas developed? Why were they considered important? What was it like to be the mathematician who created them? How did these mathematical ideas relate to the needs, concerns, values of the larger culture? This decontextualized vision inevitably discourages certain people from pursuing mathematics—those for whom the connection to the world is not a bad thing or something to be escaped; those who hunger to integrate mathematical knowledge with their lives, finding ways to apply it to real-world problems; those who, as Johnetta Cole (the president of Spelman College) describes, feel that education should serve some purpose and address a sense of social responsibility.[57]

Finally, the traditional focus on the product of mathematics, rather than the process, lends itself to a particular method of teaching—in particular, a lecture style in which the teacher is perceived as the authority figure who has the right answer, which students learn to replicate. Paulo Freire describes this as the banking model, in which the teacher "deposits" wisdom into the empty vessel of students' minds. The authority of the teacher is rarely questioned because mathematical knowledge is seen as certain, not something to be discussed or negotiated. Indeed, students are left with the impression that mathematical truths just descend from a Platonic heaven, and only the mathematical priests—professors and teachers—have access to this heaven. The most a student can learn to do is replicate perfect proofs that shape the substance of mathematics.

Certainly these issues affect men as well as women, a theme developed in Larry Copes's essay "Mathematical Orchards and the Perry Development Scheme."[58] However, it appears that these issues, in combination with other factors, can discourage women even more than men.

Women's lower levels of confidence[59] as well as their different ways of responding to authority can contribute to increased alienation from mathematics when it is presented in the traditional dualistic way described above. As Pat Rogers writes,

> The expository mode of presentation may adversely affect women more than men because it uses and appeals to authority as a means of imparting knowledge. Power in the lecture situation resides with the speaker and few students are able to cultivate their own voice. However, there are marked gender differences in relationship with authorities (Belenky et al., 1986, pp. 43–5). Males are encouraged from an early age to challenge authority and receive support for holding the floor and presenting their own views. Women, on the other hand, are encouraged to subordinate their own voice to that of authorities and are far less likely than men to maintain a sense of their own autonomy.[60]

Indeed, in their book *Women's Ways of Knowing*, Belenky et al. argue that one of the primary themes in women's intellectual and personal development is "finding their own voice." For most women, this means finding ties with personal needs, interests, or experiences. "Most of these women were not opposed to abstraction as such. They found concepts useful in making sense of their experiences, but they balked when the abstractions preceded the experiences or pushed them out entirely. Even the women who were extraordinarily adept at abstract reasoning preferred to start from personal experience."[61]

Dorothy Buerk draws similar conclusions in the context of teaching mathematics. As she discovered, for many women, connecting with mathematics means finding a way to "make it their own." She describes the process that one of the students in her seminar went through, and the difficulties she faced in making connections with mathematics. Her student wrote:

> I did not feel that I could interact with math on a personal level and therefore did not feel as though I owned or could own any of the math-

ematical ideas and concepts with which I was presented. I found math simply an exercise of going through the motions and, as Carlyle (1830, p. 965) put it, "believing for" someone else, rather than for myself. I was frustrated. . . .

I wanted to touch, to believe for myself. But I knew that interaction required process and what kind of process could there possibly be with an already finished and perfected product. It didn't matter that I had taken Calculus, or any other course for that matter. I still felt estranged from math, aggravated by it, afraid of it. . . . The first step to conquering fear is naming it. Yet how is it possible to name a thing that renders you voiceless? Unlike English class, math was not a place for ideas in process. You could not say or share something you were thinking about. You could only share with the class completed, perfected thoughts and I simply had no such thoughts concerning math.[62]

But it is difficult to find a way to tie mathematics to one's personal experience in the context of a traditional mathematics classroom, where the emphasis is on removing one's own thoughts, ideas, feelings, and impressions, and replacing them with rigorous and correct proofs, even when one is not clear how these proofs came to exist or why they are true.

In addition, women are more likely to be socially conditioned to define certain factors as high priorities in their choice of life work. For example, they may seek work that connects them to rather than isolates them from other people. Hence, they are more likely to be alienated from images of mathematics that denigrate social concerns, or that portray the physical and human world as something to be escaped.

More research is needed in order to understand the relationship between dominant visions of mathematics, and who chooses to pursue mathematics, and how gender enters into this process. However, what we do know from numerous studies is that stereotypes of mathematics as a male field, and stereotypes that women are no good at mathematics, abound. And these stereotypes can influence women's decisions not to pursue mathematics. When the teaching methods and images of mathematics further reinforce the distance that women feel in relation to mathematics, it is easy for them to just assume that the stereotypes are right, that they are indeed no good at mathematics, or it isn't for them. While men may also feel alienated from mathematics, they are less likely to see this as a personal inadequacy, and are more likely to be encouraged to pursue mathematics anyway; men do not have the background noise constantly in their head that mathematics is not an appropriate field of study for them. Thus, while this vision of mathematics can discourage both men and women, I believe that in combination with other variables, it can have an even more discouraging effect on women.

The Impact on Women Already in Mathematics

Because mathematics is seen as consisting of certain and eternal truths, and because these truths are obtained by methods that seem impeccably reliable, mathematics is often thought of as a realm of complete objectivity. Moreover, because the methods and content of mathematics are assumed to be objective, the whole discipline inherits that reputation. Consequently, any decisions that are made—even those that pertain to the human realm, such as hiring, promotions, awards, recognition, etc.—are assumed to rest on a firm foundation of objective criteria. When one looks more closely at mathematics, however, the foundation is not nearly as firm as it might seem. On many fronts, the claims to certain, eternal, and absolute knowledge can be challenged.

In reality, mathematics is a human activity that is inevitably influenced by subjective and social norms. It is not the theorems, nor even the proofs, that ultimately hinder women in mathematics; it is the misidentification of mathematics with just theorems and proofs. If the social and subjective dimension is not acknowledged, then bias against women can become invisible, hidden behind the guise of objectivity. This can happen, for example, in judgments about whether research is "important"—a judgment critical in hiring, promotion, and tenure decisions. Often people do not recognize the social component of these attitudes until they find themselves outside mainstream opinions. Lenore Blum's area of logic, which was enthusiastically supported on the East Coast, met a very different reception in California:

> I was working in a field that I felt was really terrific, and when I came to Berkeley as a post-doc, I was very young and very naive. I just assumed that everybody here would be interested in it, and I didn't understand the politics of why groups of people keep working in certain areas. And it didn't dawn on me at first, but in truth the people at Berkeley were not interested in that field at all. Not at all in model theory, and certainly not in the kind of model theory I was doing. I think it was viewed the same way as applied math was viewed in those years, as not really quite legitimate mathematics, and that logic applied to mathematics wasn't pure logic. It was the sixties, so everything had to be pure math, pure logic, etc. It is ironic to imagine that number theory or algebra would make logic impure.

What is fascinating to one person may not be fascinating to another. Nonetheless, these personal or communal judgments can be disguised as objective claims about what is important research. If someone argues, for example, that a person's work is not of seminal importance, how can this be objectively measured? Would there be universal agreement about what counts as seminal and what does not? As Einstein wrote, "My intuition was not strong enough in the field of mathematics in order to differentiate clearly the fundamentally important, that which is really basic, from the rest of the more or less dispensable erudition."[63] But if Einstein could not make these distinctions, who can?[64]

A secondary way that images of mathematics can affect women in mathematics is more subtle. There is a whole network of associations typically identified with mathematics: rational, objective, a focus on the mind. But this same set of traits is also traditionally identified with men. Moreover, the counterparts of these traits—intuitive, subjective, a focus on the body—are typically identified with women. Though women may reject these cultural associations, they are nonetheless subject to them. And because the mathematics community is a subset of society at large, they inevitably absorb these equations to some degree. As a result, women are in a more precarious position when they embrace a trait typically tied to women's sphere. One example of this is intuition. We are accustomed to hearing phrases such as "women's intuition." Yet there is no evidence that women mathematicians are any more intuitive or less logical than their male counterparts. And as we have seen, both intuition and logic are critical in doing mathematics. Nonetheless, women may be more reluctant to play up the intuitive/preformalized dimension of their work for several reasons.

As chapter 2 showed, many behaviors traditionally identified with women are perceived quite differently when women do them than when men do. If women bring children in to work, they are seen as unprofessional; if men do, it is seen as a charming anomaly. If women get angry, they are described as hysterical or irrational; if men do, they are seen as forceful in their opinions. Similarly, if women emphasize the intuitive aspect of their work, they risk being described as vague, imprecise, and non-rigorous. They are vulnerable to the stereotype that women are not mathematically inclined.

These concerns are further exacerbated by the fact that because women are often held suspect until they prove themselves, they are more reluctant to reveal the tentative early stages of their work. These stages inevitably involve mistakes and false starts, which, though seen as normal and acceptable in men, can become magnified for women, further reinforcing biased assumptions that women are no good at mathematics. Consequently, many women feel pressure to be completely certain of results before allowing them into the public arena of the mathematics community. Indeed, as Patricia Kenschaft points out, there is evidence that women are less likely to submit grants or journal articles before they are completely polished.[65]

These concerns are also complicated by other related issues. For example, it may be the case that women's ideas are more likely to be appropriated (particularly at this early, pre-publication stage) than a man's ideas, and women may be more reluctant to challenge this kind of behavior when it arises.

In addition, if it is true that women have more difficulty gaining access to the informal channels in which so much mathematical activity takes place, they have less opportunity to cultivate and develop their mathematical intuition, and to become at ease in developing and exchanging ideas at this pre-rigorous level. Yet it is in the context of this informal, intuitive exchange that so much learning and development happens. As Jerome Bruner writes, "[Intuition] is founded on a kind of combinatorial playfulness that is only possible when the consequences of error are not overpowering or sinful. Above all, it is a form of activity that depends upon confidence in the worthwhileness of the process of mathematical activity rather than upon the importance of right answers at all times."[66]

If women are less likely to feel comfortable engaging in that kind of informal exchange and interaction of ideas, they lose out on the benefits that come from it. Moreover, if, as Ravenna Helson's study indicates, women tend to have lower levels of confidence, this can contribute to their reluctance to expose the tentative stages of inquiry.

These ideas suggest questions for future investigation. What would it mean to cultivate an environment in which mathematical intuition and imagination are actively nourished? If this is a fundamental resource of mathematical research, how can we develop this resource in future and current practitioners? Such a climate is likely to have a positive impact on both men and women, but it may be particularly important for women, who are more likely to feel stifled by the prison of perfection, and whose mistakes or inevitable false starts are more likely to be magnified under the lens of doubt in the mathematics community.

This chapter is meant to bring a variety of questions to the fore. How do we think about the nature of mathematics? How does that influence how we teach mathematics, the culture of mathematics, and who goes into mathematics? This is simply an initial step in that direction. Much more work needs to be done in order to elucidate the subtle relationship between gender, visions of mathematics, and the culture of mathematics. Although these questions may be difficult to answer, I believe they can have a significant impact on the future landscape of the mathematics community. If there is one lesson that is clear from the history of mathematics, it is that questioning even the assumptions that seem most obvious and fundamental can lead to profound new ideas and visions. If we want to think hard about ways to be more inclusive of non-traditional groups, it is critical to reexamine all aspects of the discipline.

Conclusion

There are many ways to think about mathematics. At times it has been defined as the study of number, at other times as the study of shape; in this century mathematics was viewed as reducible to logic. As mathematics continues to grow in depth and breadth, it becomes increasingly difficult to capture it in a simple definition. Perhaps the most useful one was articulated by, among other people, Lynn Arthur Steen: that mathematics is the study of pattern. This includes not only patterns that arise in numbers or shapes, but also patterns in reasoning or in the natural world: in the movement of molecules, in the natural cycles of populations, or in weather patterns. Most mathematicians would specify further that mathematics is the study of patterns that are quantifiable or definable in a precise way (though some of the research on the boundaries between mathematics and physics might challenge this stipulation).

What remains unanswered, however, is the following question: Do these patterns exist "out there" independent of human beings? Do we simply observe these patterns and record them in clear and precise ways? Or are they a construction of our own minds—a way of imposing order on a random set of sensations?

I find metaphor useful in thinking about a response to these questions. If we think of reality as the stars in the sky, then mathematics is the attempt to organize those stars into patterns. Mathematics defines and gives names to those patterns. It is a powerful tool, for by grouping objects into categories, we can manipulate them in our minds and communicate ideas about them more easily. It allows us to make sense out of an enormous mass of data. And in this way it enables us to see new things, new patterns that would elude us if we saw the stars only as individual entities.

At the same time, there are many ways to impose order on the skies. Indeed, different cultures have very different images and names for the constellations: some see big dippers, bulls, or crabs, while others see gods and goddesses. These images are not inherent in the stars; they are in our mind's eye, our human attempt to define pattern. Thus, we can never know the stars in a "pure," direct sense; we know them through a lens, a lens which helps us order them so that we can label them and "make sense" out of them. Our minds are the lenses. And when one is raised to see big dippers all one's life, it is indeed quite difficult to break those habits of perception and define different patterns. In fact, it often becomes hard to see the individual stars at all; they are blurred into an entity, and it is that entity that our mind's eye becomes trained to see.

In this sense, mathematics can be thought of as the study of something "out there," and in this sense it can be seen as true or false: one cannot say there is a star where there is not. And yet, because there are many ways to impose pattern, there is not simply one mathematical reality. There are many. Moreover, because the sky is far vaster than we can ever know in its entirety, we are always viewing only pieces of it. Which piece gets focused on is influenced by individual and social factors.

For this reason, it is important to have many different people participating in mathematics. Different individuals and different cultures bring their unique perspective. They can suggest a different set of patterns, reminding us that the ones we are used to are not the only possible way to make order. They may also bring to light a different subset of the infinite set of possible stars in the sky. If mathematics becomes the practice of only one group of people, it limits the collective imagination that feeds creativity. At this point, women continue to be significantly underrepresented in mathematics, particularly at the highest echelons. Although most of the formal barriers have been dismantled, informal barriers continue to exist, though they are harder to identify. They take the form of implicit assumptions and expectations about women, about mathematicians, and about mathematics.

By examining women's lives, we see these underlying assumptions more clearly. We also see that alternative visions are possible. Their stories give us a richer sense of mathematicians' lives: that it is possible to be a woman and a mathematician, that doing mathematics is not contradictory with having children, that cooperation and collaboration can be tremendous assets to mathematical activity, that social and political concerns can be intimately connected to one's identity as a mathematician, and that mathematicians can be quite productive throughout their lives, often peaking in their fifties or sixties rather than their twenties or thirties. In this way, their stories enrich our vision not only of what it can mean to be a woman, but also of what it can mean to be a mathematician.

Ultimately, this book is meant as one step in the exploration of the culture of mathematics, and how that culture shapes the lives, choices, and work of practitioners. It can be thought of as part of a larger attempt to understand the relationship between the values and beliefs of a community, who becomes a practitioner in that community, and what kind of knowledge is produced by that community.

It is clear that women can do mathematics. The next step is to embrace them as equal members. As the famous mathematician Weierstrass said, "It is true that a mathematician, who is not somewhat of a poet, will never be a perfect mathematician."[1] It no longer seems strange for a woman to be a poet. There is no reason for it to be strange for a woman to be a mathematician.

Introduction

1. See, for example, Roberta Hall and Bernice Sandler, *The Classroom Climate: A Chilly One for Women?* or Bernice Sandler and Roberta Hall, *The Campus Climate Revisited: Chilly for Women Faculty, Administrators, and Graduate Students* (Washington, D.C.: Project on the Status and Education of Women, Association of American Colleges, 1986). Also, Patricia Kenschaft and Sandra Keith, eds., *Winning Women into Mathematics* (Washington, D.C.: Mathematical Association of America, 1991).

2. Rozsa Peter, *Playing with Infinity* (New York: Simon and Schuster, 1962), p. v.

3. I was struck by the fact that readers tended to have strong feelings about which form the profiles should take: mathematicians tended to prefer the direct interview format; non-mathematicians overwhelmingly preferred the essays. The roots of these preferences are an interesting topic in themselves. For a mathematician, knowledge takes the form of fact. Hence interviews, which seem more direct and closer to objective fact, are preferred. For those outside the sciences, the assumption is that knowledge is always embedded in interpretation. From that perspective, it seems natural and essential that I make sense of the stories, highlight the themes, and provide a larger context.

Because this book is intended for both audiences, I have preserved both forms. I am interested in knowing whether this pattern holds true with a larger pool of readers.

4. Having children violates another aspect of the ideology of a mathematician: that one must be completely focused, absorbed, even obsessed by one's work. As the famous mathematician Paul Halmos wrote, one "must love mathematics more than anything else" (Paul Halmos, *I Want to Be a Mathematician* [New York: Springer Verlag, 1985], p. 400). In such a vision, there is no room for the distraction of children and parenting. That this kind of focus is an *asset* in mathematical activity is clear, but the form such concentration takes, and that it is incompatible with having children, is not clear. Indeed, stories of women such as Mary Ellen Rudin, who was able to do mathematics in her living room while surrounded by four children, suggest that a very different model is possible.

5. Ann Hibner Koblitz, *A Convergence of Lives: Sofia Kovalevskaia—Scientist, Writer, Revolutionary* (Boston: Birkhauser, 1983).

6. Lynn Osen, *Women in Mathematics* (Cambridge, Mass.: MIT Press, 1974).

7. Kovalevskaia had gotten one unofficially. She was not allowed to attend classes and did not have an oral exam.

8. Ethel Colburn Mayne, *The Life and Letters of Anne Isabella Lady Noel Byron* (New York: Charles Scribner's, 1929), p. 477.

9. The notable exceptions, such as Kovalevskaia, Germain, Sommerville, Grace Chisolm Young, and later Emmy Noether, have been written about extensively in recent years. In addition to work by Lynn Osen, Teri Perl, and Ann Koblitz, see, for example, Louise Grinstein and Paul Campbell, eds., *Women of Mathematics: A Bibliographic Sourcebook* (New York: Greenwood Press, 1987).

10. Teri Perl, *Math Equals* (Menlo Park, Calif.: Addison Wesley, 1978).

11. Ivor Grattan-Guinness, "A Mathematical Union: William Henry and Grace Chisolm Young," *Annals of Science* 29, no. 2 (August 1972): 107.

12. See, for example, McGeorge Bundy, "Hitler and the Bomb," *New York Times Magazine,* November 13, 1988, and letter to editor by Richard Wolfson, ibid., April 1989.

13. National Research Council, Commission on Human Resources, *Climbing the Academic Ladder: Doctoral Women Scientists in Academe* (Washington, D.C.: National Academy of Sciences, 1979), p. 20.

14. Lucy Sells, "The Mathematics Filter and the Education of Women and Minorities," in L. H. Fox, L. Brody, and D. Tobin, eds., *Women and the Mathematical Mystique* (Baltimore: Johns Hopkins University Press, 1980).

15. "Foreword," in Susan Chipman, Lorelei Brush, and Donna Wilson, eds., *Women and Mathematics: Balancing the Equation* (Hillsdale, N.J.: Lawrence Erlbaum, 1985).

16. A particularly useful map of the territory is provided in Gilah Leder's article "Gender Differences in Mathematics: An Overview," in *Mathematics and Gender,* ed. Elizabeth Fennema and Gilah Leder (New York: Teachers College Press, 1990).

17. See, for example, Fennema and Leder, *Mathematics and Gender;* Sue Rosser, *Female-Friendly Science* (New York: Pergamon Press, 1990); Myra Sadker and David M. Sadker, *Failing at Fairness* (New York: Charles Scribner's Sons, 1994); "The AAUW Report: How Schools Shortchange Girls" (AAUW Educational Foundation and National Education Association, 1992); Catherine Krupnick, "Women and Men in the Classroom: Inequity and Its Remedies," *Teaching and Learning: Journal of the Harvard Danforth Center* 1, no. 1 (May 1985).

18. Fennema and Leder, *Mathematics and Gender.*

19. Rosser, *Female-Friendly Science,* p. 69. Also Fennema and Leder, *Mathematics and Gender.*

20. Fennema and Leder, *Mathematics and Gender,* chap. 5.

21. Lynn Billard, "The Past, Present, and Future of Academic Women in the Mathematical Sciences," AMS *Notices* 38, no. 7 (September 1991): 708: "There were 9,259 women and 14,454 men in 1972, and 5,006 women and 6,593 men in 1982 receiving the bachelors degree in mathematics."

22. *Notices* 40, no. 6 (July/August 1993): 608.

23. Ibid., p. 1167.

24. Ibid., p. 603. Women received 14.6 percent of the doctorates granted between 1980 and 1990 (ibid. 38, no. 6 [September 1991]: 718).

25. Ibid. 40, no. 6 (July/August 1993): 605.

26. Ravenna Helson, "Women Mathematicians and the Creative Personality," in *Readings on the Psychology of Women,* ed. Judith Bardwick (New York: Harper and Row, 1972). Originally printed in *Journal of Consulting and Clinical Psychology* 36, no. 2 (1971): 210–220.

1. Rugged Individualism and the Mathematical Marlboro Man

1. Donald J. Albers, "Freeman Dyson: Mathematician, Physicist, and Writer," *College Mathematics Journal* 25, no. 1 (January 1994): 13.

2. D. E. Smith, "Dinner in Honor of Professor David Eugene Smith," *Mathematics Teacher* 19 (May 1926): 279.

3. "The spirit of adventure, a sort of heroism, animates the mathematician far

more than his formulas." Quote by Gonseth from François Le Lioinnais, in "Beauty in Mathematics," in *Mathematics: People, Problems, Results,* ed. Douglas Campbell and John Higgins (Belmont, Calif.: Wadsworth, 1984), vol. 3, p. 87.

4. Serge Corrado, "On Some Tendencies in Geometric Investigations," *Bulletin of the American Mathematics Society* 10 (June 1904): 453. In a similar vein, the mathematician Bolyai wrote to his son, who was seeking to answer the same mathematical question about the Fifth Postulate that his father had spent most of his life on: "I have traveled past all the reefs of the infernal Dead Sea and have always come back with a broken mast and a torn sail." Fortunately his son's quest was not in vain; he was able to solve the problem that had eluded his father for decades. As he wrote a short while later: "Out of nothing I have created a new and wonderful world!"

5. James Pierpont, "Mathematical Rigor, Past and Present," *Bulletin of the American Mathematics Society* 34 (January–February 1928).

6. Samuel Jones, *Mathematical Wrinkles* (Gunter, Tex.: Samuel I. Jones, 1912), p. 257.

7. Ethel Coburn Mayne, *The Life and Letters of Anne Isabella Lady Noel Byron* (New York: Charles Scribner's Sons, 1929), p. 477.

8. As James R. Newman writes in *The World of Mathematics* (New York: Simon and Schuster, 1956): "The mathematician is still regarded as the hermit who knows little of the ways of life outside his cell, who spends his time compounding incredible and incomprehensible theorems in a strange, clipped, unintelligible jargon."

9. Camille Jordan, "Notice sur la vie et les travaux de Camille Jordan," *L'enseignement mathematique* 3, ser. 2 (April–June 1957): 92.

10. Donald Weidman, "Emotional Perils of Mathematics," *Science* 149 (September 1965): 1048.

11. Gina Kolata, "Math Whiz Who Battled 350-Year-Old Problem," *New York Times,* June 29, 1993, pp. C1, 11.

12. In this chapter I use the term *autonomy* interchangeably with *independence, self-sufficiency,* or *individualism.* Philosophers, however, make distinctions between these terms; autonomy is often equated with "self-rule," which is different from self-sufficiency, for example. Self-rule implies an ability to define one's own guidelines, or laws of action, while self-sufficiency implies that one does not need anyone else to sustain oneself. In the context of these distinctions, I would argue that independence and self-sufficiency are the central values identified with the image of a mathematician. However, one of the functions of these traits is preserving a certain amount of autonomy (the capacity for self-rule), which is seen as important for the creative work of a mathematician. I would also argue that isolation is seen as one way (perhaps an extreme way) of ensuring independence.

13. Donald Albers, Gerald Alexanderson, and Constance Reid, eds., *More Mathematical People* (Boston: Harcourt Brace Jovanovich, 1990).

14. Elizabeth Fennema and Gilah Leder, eds., *Mathematics and Gender* (New York: Teachers College Press, 1990).

15. For example, by arranging to have them give talks at professional meetings, and by facilitating connections with colleagues.

16. Institutional affiliation is another key factor. Uhlenbeck worries that students who work with her at the University of Texas, Austin, may be handicapped in future job searches, both because she feels in some ways like an outsider to the

mainstream math community and because it is harder to get a job at a place such as UT than at Harvard or Princeton.

17. This point was established quite dramatically in, for example, Michele Paludi and William Bauer, "Goldberg Revisited: What's in an Author's Name?" *Sex Roles* 9, no. 3 (1983), in which they showed that the same article is given a higher rating when the author has a man's name than when the author has a woman's name. Other work on how male and female academics are perceived by both colleagues and students includes Susan Basow and Nancy Silberg, "Student Evaluations of College Professors: Are Female and Male Professors Rated Differently?" *Journal of Educational Psychology* 79, no. 3 (1987): 308–314; Nadya Aisenberg and Mona Harrington, *Women of Academe* (Amherst: University of Massachusetts Press, 1988); Julie Ehrhart and Bernice Sandler, *Looking for More Than a Few Good Women in Traditionally Male Fields* (Washington, D.C.: Project on the Status and Education of Women, Association of American Colleges, 1987); and Patricia Kenschaft and Sandra Keith, *Winning Women into Mathematics* (Washington, D.C.: Mathematical Association of America, 1991).

18. This point is developed more fully in chapter 2 as well as in several of the profiles in this book. In addition, see, for example, Bernice Sandler and Roberta Hall, *The Campus Climate Revisited: Chilly for Women Faculty, Administrators, and Graduate Students* (Washington, D.C.: Project on the Status and Education of Women, Association of American Colleges, 1986); Kenschaft and Keith, *Winning Women into Mathematics;* or *Barriers to Equality in Academia: Women in Computer Science at MIT,* report prepared by female graduate students and research staff of the Laboratory for Computer Science and the Artificial Intelligence Laboratory at MIT, February 1983.

19. Gina Kolata, "Hitting the High Spots of Computer Theory," *New York Times,* December 13, 1994, pp. C1, C10.

20. For Mary Ellen Rudin, it was initially the Moore family that she joined during graduate school, but over time it became the community that she helped shape and for which she became a center at the University of Wisconsin. This community is the one Judy Roitman connected with and which has provided the intellectual stimulation that has fed her own work. For Fan Chung there has been a natural community, first at Bell Labs and then at Bellcore, and her husband became a constant and important source of collaboration and stimulation. For Joan Birman it was Ralph Fox and the group she connected with at Princeton. For Karen Uhlenbeck, Yau and the colleagues she met at Princeton played this role. Even though Lenore Blum maintained ties to the Berkeley mathematics community even while she was teaching at Mills College, it was critical for her to find a way to reconnect and actively engage with a mathematics community in order to return to research after having devoted years to teaching and forming programs and networks for women in math. Such a reentry into research is almost unheard of. There were certainly no models she knew of to follow, so she had to create a path of her own. For Blum this entailed commuting to the East Coast to reconnect with the mathematics community there and reestablish her mathematical reputation. This turned out to be a successful strategy. She developed ties with researchers at MIT and IBM, and ultimately won an NSF fellowship for women, which was tremendously valuable in her gaining public recognition and support in her return to research.

21. "Empirical studies have reported that professional connections and collegial interaction are associated with productivity in science. In an analysis of hard

sciences in 18 disciplinary areas, Biglan (1973) found that level of social connectedness (measured as numbers of collaborators, numbers of persons worked with on research, teaching, and administration, and sources of influence on research goals and teaching procedures) was related to publication productivity. Similarly, with a sample of faculty in a private research university and two private colleges, Findelstein (1982) reports that faculty with strong collegial ties both on and off campus had the highest publication rates." Mary Frank Fox, in Harriet Zuckerman, Jonathan Cole, and John Bruer, eds., *The Outer Circle* (New York: Norton, 1991), p. 196.

22. Most of the richest collaborations involve a complex relationship in which the lines between professional and personal blur. As Joan Birman explains, if you really like the collaborator as a friend, the collaboration can be much richer. For women, that intimacy of friendship is often easier to develop with other women for many reasons: they understand what it is like to be a women in mathematics; it can fill a longing for female companionship, which is rare in mathematical research; the sexual politics tend to be less loaded, so it can be easier for women to be completely at ease; and there is less of a feeling of having to prove oneself. This was all true for Birman in her collaboration with Carolyn Series, which she describes as one of her best collaborations. "She is the only woman [I collaborated with]. Certainly the fact that we were both women had a big impact on that. We worked together very well."

23. See, for example, Lynne Billard, "The Past, Present, and Future of Academic Women in the Mathematical Sciences," AMS *Notices* 38, no. 7 (September 1991), or Sandler and Hall, *The Campus Climate Revisited.*

24. Lenore Blum provides an interesting case study of this phenomenon. For a deeper examination of how and why this can happen, see, for example, Aisenberg and Harrington, *Women of Academe,* or Zuckerman, Cole, and Bruer, *The Outer Circle.*

25. J. L. Synge, *Kandelman's Krim* (London: Jonathan Cape, 1957), p. 84.

26. Gabor Szego describes the young men in Hungary who were inexorably drawn to mathematics through the problems in the *Hungarian Mathematics Journal:* "They were bound finally and unalterably to the jealous mistress that mathematics is." Gabor Szego, *Hungarian Problem Book I,* comp. Jozsef Kurschak (Washington, D.C.: Mathematical Association of America, 1963), p. 7. Paul Halmos, in his "automathography," says that "to be a mathematician, you must love mathematics more than anything else, more than family, more than religion, more than any other interest" (*I Want to Be a Mathematician* [New York: Springer Verlag, 1985], p. 400). This theme is echoed by the famous mathematician David Hilbert, who wrote that it was mathematics—which they "loved more than anything else"—that brought him and Felix Klein together.

27. Along these lines, it is important to engage the question of what is meant by cooperation in mathematics. For some, it simply means working with someone on a problem. This could involve working separately and coming together occasionally to report results. An analogy might be that of two people climbing a mountain separately, hoping to meet at the top. Some mathematicians, such as Fan Chung, have explicitly articulated guidelines that are useful in collaborative relationships—see her profile in chapter 2. Other scholars have written extensively about ingredients for effective cooperative group work. Johnson and Johnson, for example, have developed these ideas in the context of education. They argue that effective cooperation in groups occurs when both group goals and individual

accountability exist. Moreover, specific social skills need to be developed in order to work with the kinds of issues that are likely to arise in group interaction. In addition, time must be set aside to address the process itself, and to ensure a continued productive relationship between all members of the group. See, for example, David W. Johnson, Roger T. Johnson, and Edythe J. Holubec, *Circles of Learning* (Edina, Minn.: Interaction Book Company, 1993), or Neil Davidson, ed., *Cooperative Learning in Mathematics* (Menlo Park, Calif.: Addison Wesley, 1990).

28. In a similar way, people have varying degrees of connection with the mathematical community. Some attend many meetings, are in contact with a wide range of colleagues, and have a regular exchange of ideas. Others rarely have the opportunity to make contact beyond their immediate environment, or have not developed close ties with a community of mathematicians in their field.

29. More specifically, her field of research is geometric, non-linear, partial differential equations.

30. As Karen related in a later discussion: "It is interesting that the one really good professional experience I recall from those days is the fellowship from the University of Illinois I received. I was a given a semester off from teaching, and spent a good part of it in New York, living with my husband's parents and occupying an office at the Courant Institute. It was very inspiring. What went wrong is that I was too ambitious. Abe Taub, at Berkeley, told me about some of the hard problems in understanding shock waves, and I pursued this lead, with no background or co-worker. They were/are beautiful geometric problems. It was years before people with the proper background and vision made a dent in them."

31. As Uhlenbeck goes on to say, "People who live lives like mine are not looking for that sort of thing. It's like you did everything you could so that you wouldn't be in the center of things with people looking up to you."

32. Radcliffe, an all-women's college, was closely associated with Harvard, and Radcliffe women took classes at Harvard. But their numbers at that time were still quite small, and their dorms, etc., were physically isolated from Harvard's center. In this sense, Radcliffe women were seen more as visitors than as traditional Harvard students.

33. Marian Boykan Pour-El, "Spatial Separation in Family Life: A Mathematician's Choice," in *Mathematics Tomorrow,* ed. Lynn Arthur Steen (New York: Springer Verlag, 1981), p. 188.

34. Ibid.

35. On her choice of courses, she said, "I always liked math because I thought I could understand a lot about the universe by understanding mathematics and physical models, and I wanted biology simply because I wanted to understand the biological human race in a sense." In fact, because there was insufficient opportunity to study everything she was interested in as an undergraduate, she devised a plan, "I had the idea that I would go to graduate school and then go back to college again to major in other things that I liked. In other words, this time around in math, next time around in biology."

36. Pour-El, "Spatial Separation in Family Life," p. 188. In the interview for this book she says, "I had very good friends even though we were very different. They discussed boys and dates. I was too busy riding subways. They knew I wasn't dating, but we were still good friends. During that period I realized unconsciously that you could do anything you really wanted and still have friends."

37. Pour-El, "Spatial Separation in Family Life," p. 189.

38. She never applied for the job. Penn State approached Pour-El asking if she

would be interested in a position there. As was almost always the case, these decisions and contacts happened through informal channels, and took place behind the scenes.

2. What's a Nice Girl Like You Doing in a Place Like This?

1. The cartoon is from Patricia Kenschaft and Sandra Keith, eds., *Winning Women into Mathematics* (Washington, D.C.: Mathematical Association of America, 1991).

2. This is the title of a very interesting article by Leone Burton on this topic: "Femmes et mathematiques: Y a-t-il une intersection?" *AWM Newsletter* 18 (November–December 1988): 17.

3. As Plato wrote in the *Symposium,*

> Those whose creative instinct is physical have recourse to women, and show their love in this way, believing that by begetting children they can secure for themselves an immortal and blessed memory hereafter forever; but there are some whose creative desire is for the soul, and who long to beget spiritually, not physically, the progeny which it is the nature of the soul to create and bring to birth.

He goes on to say, "Everyone would prefer children such as these [from the soul] to children after the flesh."

4. Paraphrase from Immanuel Kant's *Observations on the Feeling of the Beautiful and the Sublime* (Berkeley: University of California Press, 1960), pp. 78–79.

5. Fewer than ten women are typically cited for a period of two thousand years—see, e.g., Lynn M. Osen, *Women in Mathematics* (Cambridge, Mass.: MIT Press, 1974), or Teri Perl, *Math Equals* (Menlo Park, Calif.: Addison Wesley, 1978).

6. Judy Roitman says, "More sorts of things include not being taken seriously or being somewhat invisible in conversations, having passes made at you by professors, and having the feeling that they were only putting up with you so that they could do that—generally being asked questions which typecast you into certain roles, for example, when my advisor said, 'I thought you were going to teach junior high,' because I had done it over the summer to earn some money. . . . Just the general feeling that you were not being taken seriously."

7. Later Marian Pour-El describes how being a woman complicated her relationship with fellow classmates in graduate school in specific practical ways: "Many times we decided to go to a movie or a concert, so I would go along too, and I always wanted to pay my own way because I wanted to be able to suggest a movie too. But they couldn't accept that. They would always try to be a pace ahead of me, to pay my way before I got there. There was no such thing as 'dutch' then. It was considered an insult [for them not to pay] because I was a woman. I told them that I didn't want to be considered a 'woman.' They couldn't understand that either."

8. G. H. Hardy, *A Mathematician's Apology* (Cambridge: Cambridge University Press, 1967), p. 77.

9. Donald Albers, Gerald Alexanderson, and Constance Reid, *More Mathematical People* (Boston: Harcourt Brace Jovanovich, 1990), p. 233.

10. Ibid.

11. Nadya Aisenberg and Mona Harrington, *Women of Academe* (Amherst: Univer-

sity of Massachusetts Press, 1988); Elaine Martin, "Power and Authority in the Classroom: Sexist Stereotypes in Teaching Evaluations," *Signs* 9, no. 3 (1984).

12. Susan Basow and Nancy Silberg, "Student Evaluations of College Professors: Are Female and Male Professors Rated Differently?" *Journal of Educational Psychology* 79, no. 3 (1987): 308–314; or Susan Basow, "Effects of Teacher Expressiveness: Mediated by Teacher Sex-Typing," *Journal of Educational Psychology* 82, no. 3 (1990): 599–602.

13. Bernice Sandler and Roberta Hall, *The Campus Climate Revisited: Chilly for Women Faculty, Administrators, and Graduate Students* (Washington, D.C.: Project on the Status and Education of Women, Association of American Colleges, 1986).

14. G. B. Kolata, "Cathleen Morawetz: The Mathematics of Waves," *Science* 206 (October 12, 1979): 207.

15. See, for example, "The AAUW Report: How Schools Shortchange Girls" (AAUW Educational Foundation and National Education Association, 1992); Roberta Hall and Bernice Sandler, *The Classroom Climate: A Chilly One for Women?* (Washington, D.C.: Project on the Status and Education of Women, Association of American Colleges, 1986); Myra Sadker and David M. Sadker, *Failing at Fairness* (New York: Charles Scribner's Sons, 1994).

16. This is an issue that is much less likely to arise for a man than for a woman. Even in the unlikely case, for example, where a man is in the position of having a female advisor, if the personal dynamics are problematic, he is likely to have many alternative choices of advisors of the same sex. For women, this option is typically nonexistent.

17. See, for example, Elizabeth Fennema and Gilah Leder, eds., *Mathematics and Gender* (New York: Teachers College Press, 1990), or Sadker and Sadker, *Failing at Fairness*.

18. See, for example, Mary Belenky, Blythe Clinchy, Nancy Goldberger, and Jill Tarule, *Women's Ways of Knowing* (New York: Basic Books, 1986), as well as these interviews.

19. Sadker and Sadker, *Failing at Fairness*. See also their article "Sexism in the Classroom: From Grade School to Graduate School," reprinted in *AWM Newsletter* 20, no. 6 (November– December 1990).

20. See Patricia Kenschaft and Sandra Keith, eds., *Winning Women into Mathematics* (Washington, D.C.: Mathematical Association of America, 1991).

21. It should be noted that Marcia Groszek has since had a child and worked out an arrangement whereby her husband is the primary caregiver.

22. Julie Ehrhart and Bernice Sandler, *Looking for More Than a Few Good Women in Traditionally Male Fields* (Washington, D.C.: Project on the Status and Education of Women, Association of American Colleges, 1987).

23. Sue Geller says, "I've been told by a number of (male) editors [at the less prestigious journals] that they have found their best papers written by women. In many cases they asked, 'Why didn't [these women] submit to more prestigious journals?'" (*Science* 255 [March 1992]: 1383). See also Kenschaft and Keith, *Winning Women into Mathematics*.

24. Alfred North Whitehead, "The Nature of Mathematics," in Whitehead, *An Introduction to Mathematics* (London: Oxford University Press, 1948), p. 2.

25. Marian Boykan Pour-El, "Spatial Separation in Family Life: A Mathematician's Choice," in *Mathematics Tomorrow*, ed. Lynn Arthur Steen (New York: Springer Verlag, 1981).

26. When I first interviewed Mary Ellen Rudin in 1988, an interview with her had recently been published, so I tried not to overlap too much with the material covered there. As a result, many of the excerpts in this essay come from that previous interview, which was later published in Albers, Alexanderson, and Reid, *More Mathematical People.*

27. Albers, Alexanderson, and Reid, *More Mathematical People,* p. 296.

28. Ibid., p. 302.

29. Ibid., p. 286. Mary Ellen had one sibling, a brother who was about ten years younger than she. So although she was not an only child, that is in many ways how she felt growing up.

30. Ibid., p. 288.

31. Ibid.

32. Ibid., p. 289.

33. Ibid., p. 292.

34. Ibid., p. 290.

35. Ibid., p. 296.

36. Mary Ellen argues that in a way, each of her children displays a certain kind of mathematical talent:

> Our oldest daughter is a linguist. She is an expert on syntax, Bulgarian syntax in particular, but she knows many other languages—south Slavic languages are her specialty. She's married to an anthropologist. Our second daughter is an engineer. She got an undergraduate degree in physics. She works for 3M in Minneapolis and is married to a computer scientist and engineer. Our youngest son is going to be a biochemist and an M.D. He's still in school. And our retarded son is the janitor for the local pizza parlor. Walter says he's our greatest success. He's living way beyond his intelligence, while the others are just living up to theirs! I claim that all these kids have inherited the family talent. They're all mathematicians of a sort. . . .
>
> How does mathematical talent show itself? It's in pattern recognition. And the linguist daughter, the engineering daughter, the geneticist—it's obvious with them. But even our retarded son has a tremendous amount of this ability. He doesn't have very good judgment, but he has certain specialized talents which seem to me to be very much of the pattern recognition type. He loves history. He can tell you what happened on certain dates. He doesn't know what the facts mean, but he likes to fit them together. He also knows the bus system in Madison absolutely perfectly. If you want to go from anywhere to anywhere at a certain time, he can tell you when the bus will come and where it will go. (Ibid., p. 135)

37. This interview was conducted in June 1989 and updated in July 1996.

38. Fan had a good experience with her advisor, calling him "very kind and helpful." At the time she took this positive relationship for granted, but over the years she has heard of many incidents in this country where advisors take advantage of their students or claim credit for their work. Because in China so much of one's status and prestige is tied to being a virtuous person, it was shocking for Fan to realize that such integrity was not universal among advisors. It is interesting to note that Herbert Wilf had many women graduate students.

39. The term *concrete* comes from combining "continuous" and "discrete" mathematics.

40. Fan has strong ideas about current mathematics education, and in particular some of the existing deficiencies that exist in exposing students to fields such as combinatorics or discrete mathematics, which have clear applications:

> We are in the midst of the rapid development of not just discrete mathematics but in using different tools in different areas of mathematics. Nowadays, the problems are much more complicated, and a large amount of information is being exchanged. Bigger and bigger computers are being built. I hate to talk about applications because mathematicians usually want to stay away from that, but the fact is that the technology is moving so fast: computer chips are becoming smaller and smaller, the capacity of the chip doubles every year in the past ten years, the capacity and performance of fiber doubles every year. But look at the university and the courses they offer at the university. In many cases, the same notes can be used year after year, while in the meantime, technology is flying ahead. The number of students majoring in mathematics is declining. It used to be a healthy number, and now there are relatively few. How can you convince a bright young student to do mathematics when all the exciting things are happening outside the curriculum? That's a very serious problem. Not everybody should be mathematicians, but a good number of them with mathematical talent and interest should probably be encouraged. Why should they be convinced to take on such a field? I was looking for young mathematicians to bring in here, ones who have a broader background, who know something about what's going on, who know computer science, and the mathematics side, who can take advantage of what's going on now, and at least appreciate the power of computing. But it's relatively hard to find good mathematicians with this background. They are well paid here, but it's hard to find them. Most places should at least provide the students with options. A lot of material in this concrete mathematics course is not available to students, and that deprives them of options. It's the edge they need for what they are going to face later on when they set out in their jobs. I'm not saying that they should all learn discrete math. I am saying that they should at least be aware of it and have the option to take such courses.

In updating this interview, Fan felt that things have improved significantly in this area.

3. Is Mathematics a Young Man's Game?

1. G. H. Hardy, *A Mathematician's Apology* (Cambridge: Cambridge University Press, 1967), p. 63.

2. Alfred Adler, "Mathematics and Creativity," in *Mathematics: People, Problems, Results*, vol. 2 (Belmont, Calif.: Wadsworth, 1984), p. 5.

3. Andre Weil, "The Future of Mathematics," *American Mathematical Monthly* 57 (May 1950): 296.

4. Sylvia Wiegand, "Grace Chisolm Young," *Association for Women in Mathematics Newsletter* 7 (May–June 1977): 6.

5. See, for example, Jonathan Cole and Burton Singer, "A Theory of Limited Differences: Explaining the Productivity Puzzle in Science," in *The Outer Circle,* ed. Harriet Zuckerman, Jonathan Cole, and John Bruer (New York: Norton, 1991). This article focuses primarily on gender differences in cumulative productivity.

Nancy Stern, in "Age and Achievement in Mathematics: A Case-Study in the Sociology of Science," in *Social Studies of Science,* vol. 8 (London: Sage, 1978), pp. 127–140, cites Cole's study "Age and Scientific Performance" (unpublished paper, SUNY at Stony Brook, 1976) and Wayne Dennis, "Age and Productivity among Scientists," *Science* 123 (1956): 724.

6. See note 5.

7. Though this may not correlate exactly with what we mean by quantity of research, it is certainly a reasonable measure of it.

8. Stern, "Age and Achievement in Mathematics," Table 2, p. 134.

9. The mean number of singly authored papers published during this five-year period (1970–74) for these age categories was 3.27, 3.97, 3.24, 2.37, 2.16, and 3.43, respectively, while the mean number of co-authored papers was 1.73, 3.36, 2.94, 1.13, 3.03, and 2.69. See Stern, "Age and Achievement in Mathematics," p. 134. Thus mathematicians sixty and over were publishing, on average, more singly authored papers than those under thirty-five.

10. She gives as an example David Hilbert's proof of a general finiteness theorem, which essentially killed the subject of invariant theory.

11. She analyzes the number of citations of leading mathematicians, in particular those who are members of the National Academy of Sciences, and compares them with a large sample of mathematicians at some of the most prestigious universities in the United States. While the mathematicians in the National Academy of Sciences do not have a higher number of *publications* on average than the comparison group, the number of *citations* for their work is nearly double that for the comparison mathematicians. Thus there does seem to be at least a reasonably strong correlation between significance of work, as recognized by the general scientific community, and the number of citations.

Many other measures might be used to evaluate mathematical development or productivity: the number of students one has had, the number of ideas contributed to other people's research, quickness, the ability to develop new theories, connections made between different fields, or important questions posed. As far as this author knows, no studies have been done to determine the correlation between age and productivity with respect to these other measures.

12. Stern, "Age and Achievement in Mathematics," p. 135.

13. Jenny Harrison did an informal survey with respect to the age at which mathematicians do their best work. While men say it is between the ages of twenty-five and forty, women say it is between the ages of thirty-five and fifty, a difference of ten years.

14. Zuckerman, Cole, and Bruer, *The Outer Circle,* p. 17.

15. D. E. Smith, *The Teaching of Geometry* (Boston: Ginn, 1911), p. 333.

16. Bela Bollobas, *Littlewood's Miscellany* (Cambridge: Cambridge University Press, 1986), p. 15.

17. See, e.g., Robert Edouard Moritz, ed., *Memorabilia Mathematica* (Washington, D.C.: Mathematical Association of America, 1942), p. 123.

18. For an interesting discussion of this issue in science more generally, see Mary

Frank Fox, "Gender, Environmental Milieu, and Productivity in Science," in Zuckerman, Cole, and Bruer, *The Outer Circle*.

4. Women and Gender Politics

1. These words are used by Gian-Carlo Rota to describe the work of the mathematician Stan Ulam. "The Lost Cafe," in *From Cardinals to Chaos*, ed. Necia Grant Cooper (Cambridge: Cambridge University Press, 1988), p. 31.

2. Mary Ellen's views on this topic do seem to have evolved over time. First, she recognizes more clearly that the issues women face today are quite different from the issues she faced early in her life:

> In my time it was not a problem simply because I didn't have a career and I wasn't teaching full-time and I didn't have any responsibilities in math-ematics. I just had the nice things: somebody to talk to, graduate students, seminars, people who love your research. And it didn't make any difference that I wasn't being paid, because I had support. I didn't feel guilty about it. Every woman I know now would feel horribly guilty about that—my daughters and everybody else. But I didn't have that problem. I was doing it for fun, as an amateur. It's a luxury that nobody can afford these days.
>
> If I were beginning today, I would do exactly what everybody else is doing, I'm sure. I would go and get a job somewhere. It would be a teaching job that would be full-time, and I would have to struggle to do the research, and I probably wouldn't do as much as I did simply because I wouldn't have the time to. I probably would have fewer children than I had because most women today do. But you do things in the time in which you are born, and I did things in the time in which I was born.

Second, while she still feels that focusing on the problems can in some sense make them a bigger deal than they need to be, she also acknowledges that people's consciousness has been raised, and some of the behavior that was once seen as acceptable is no longer viewed that way. As she says, "Men try not to do things that they might have done another time without thinking about it." She draws an analogy to smoking:

> Smoking is not good for people and in general, I think it's a bad idea. These days, if someone lights up a cigarette, practically everyone in the room points a finger at them as if to say no, no, no, bad thing, you're offending me. And that would not have been done at all, once upon a time, and it isn't done anywhere in the world except here. But it has helped a lot of people to quit smoking to do that. Now, then, it's helped a lot of people who were doing things that were offensive to women at one time, to have their consciousnesses raised, as the saying goes. I think it's basically a good thing. Sometimes it's carried to extremes, sometimes there are women who are overly sensitive. Anything that you do that changes things, you have to overdo it a little bit in order to make people conscious of it. I don't want to devote my life to it, though.

3. The intersection of gender politics and mathematics is a subset of the larger question of how mathematics relates to politics more generally. Why is there such a desire to keep mathematics and politics separate? In part this arises from the belief that mathematics is supposed to be able to shed social fingerprints, giving us access to truth in a pure sense. In this vision, politics is seen as irrelevant. But the desire to keep mathematics and politics separate emerges for other reasons as well. When the destructive potential of mathematical and scientific research became undeniable with the development of atomic weapons, many scientists and mathematicians actively wanted to divorce themselves from political causes. By going deeper into "pure" research and disassociating themselves from "applied" mathematics, they could maintain some vestige of the idea that their work was simply a quest for knowledge (for knowledge's sake) rather than a means of attaining some utilitarian goal. In this way, they could take comfort in the words of the famous English mathematician G. H. Hardy: "There is one comforting conclusion which is easy for a real mathematician. Real mathematics has no effects on war. . . . So a real mathematician has his conscience clear. . . . Mathematics is . . . a harmless and innocent occupation." G. H. Hardy, *A Mathematician's Apology* (Cambridge: Cambridge University Press, 1985), p. 140.

4. Teri Perl, *Women and Numbers* (San Carlos, Calif.: Wide World Publishing/ Tetra, 1994).

5. A phrase originally used by Mary Gray, the first president of the AWM.

6. Julia Robinson claimed that she liked her marginal status because health problems would have prevented her from fulfilling the responsibilities of a full-time professor. But as Lenore and others have pointed out, it is hard to know if Julia accepted this arrangement because that was what was available and she wanted to make the best of it, or if she would rather have had a position commensurate with her mathematical stature if it had been offered to her.

7. The seminar was organized by Steve Smale, Moe Hirsch, and John Rhodes.

8. The AWM was founded in 1971. Mary Gray was the first president, from 1971 to 1973, and Alice Schafer the second, from 1973 to 1975.

9. Perl, *Women and Numbers.*

10. In a break with traditional administrative structures, Lenore co-chaired the department with her colleague Diane MacIntyre for six years. Lenore particularly enjoyed this relationship because it brought twice as much creative energy to the job; indeed, they became like partners (a relationship that Lenore has formed with several other women who have also been very significant in her life). They discovered that being co-chairs gave them more power than working individually, since they could always argue that they had to check with each other before making any big decisions. As Lenore said, it is harder to intimidate two people than one.

11. This phrase was first coined by Lucy Sells in the early 1970s.

12. Variations of this model are now much more popular. Elements are incorporated into the system developed by Uri Treisman to attract and maintain students of color to mathematics.

5. Double Jeopardy: Gender and Race

1. According to Patricia Kenschaft and Sandra Keith, eds., *Winning Women into Mathematics* (Washington, D.C.: Mathematical Association of America, 1991), of

the new doctorates in 1988–89, only 7 out of 904 were granted to black women (U.S. citizens), 4 to Hispanic women (U.S. citizens), and none to Native American women. Thus, these minority women accounted for less than 1 percent of the total number of doctorates awarded in mathematics in the U.S. that year.

2. AMS *Notices* 37, no. 5 (May/June 1990).

3. Vivienne Malone-Mayes died in 1995. This essay is written in her memory, with deep appreciation for the work she did and gratitude for the doors she has opened for other women of color.

4. The first white woman to receive a Ph.D. in mathematics in the U.S. did so in 1886, but no black woman received a Ph.D. until 1949. In that year two received their degrees simultaneously: Evelyn Boyd Granville (Vivienne Malone-Mayes's teacher at Fisk), from Yale, and Marjorie Lee Brown, from the University of Michigan. No other black women held doctorates in mathematics when Malone-Mayes began graduate school in 1961.

5. Vivienne Malone-Mayes, "Black and Female," *Association for Women in Mathematics Newsletter* 5, no. 6 (1975).

6. Constance Carroll, "Three's a Crowd: The Dilemma of the Black Woman in Higher Education," in *All the Women Are White, All the Blacks Are Men, but Some of Us Are Brave,* ed. Gloria Hull, Patricia Bell Scott, and Barbara Smith (New York: The Feminist Press, 1982), p. 120.

7. At the time of the interview, she had a great deal of difficulty walking, had gone through chemotherapy, had serious blood problems, and more than once had come close to dying. In a downward spiral, she began taking medication to help her deal with these problems, leading to further complications.

8. New York University and Courant had separate fellowships, though technically Courant is a part of NYU.

6. The Quest for Certain and Eternal Knowledge

1. "Mathematics: Quantity and Order," in J. G. Crowther, ed., *Science Today* (London: Eyre and Spottiswoode, 1934), p. 297.

2. Leone Burton in her article "Femmes et mathematiques: Y a-t-il une intersection?" formulated a related description of mathematics. As she put it, "Three shifts can be detected over time in the understanding of mathematics itself. One is a shift from completeness to incompleteness, another from certainty to conjecture, and a third from absolutism to relativity." Her article has an interesting discussion of these ideas and their relationship to gender. See *AWM Newsletter* 18 (November–December 1988): 17.

3. Albert Einstein, "Autobiographical Notes," in Paul A. Schilpp, ed., *Albert Einstein: Philosopher-Scientist* (New York: Tudor, 1951), p. 9.

4. J. L. Synge, *Kandelman's Krim* (London: Jonathan Cape, 1957), p. 13.

5. Bertrand Russell, "Reflections on My Eightieth Birthday," in Russell, *Portraits from Memory* (New York: Simon and Schuster, 1956).

6. G. H. Hardy, *A Mathematician's Apology* (Cambridge: Cambridge University Press, 1985), p. 76.

7. Bertrand Russell, *My Philosophical Development* (London: Routledge, 1993).

8. Cassius J. Keyser, *The Human Worth of Rigorous Thinking* (New York: Scripta Mathematica, 1940), p. 48. One way in which the idea that mathematics is eternal knowledge gets conveyed is in how the history of mathematics is presented: a linear progression of truths, each built on, and generalizing, the truths that came

before. It rests on the belief that a mathematical theorem is never later proven to be false. At worst, it may turn out to be an incomplete idea. In such a case, however, it is assumed that the idea can be expanded to a larger, more accurate abstraction which then becomes the "real" truth. This allows us to maintain our belief in the immortality of the original idea. Though some scholars have begun to question this model, it is not easily altered. In the section that follows I begin to examine why.

9. Philip Davis and Reuben Hersh, *The Mathematical Experience* (Boston: Houghton Mifflin, 1982), p. 34.

10. E. P. Northrop, "Mathematics in a Liberal Education," *American Mathematical Monthly* 52 (March 1945): 133.

11. Davis and Hersh, *The Mathematical Experience*, p. 318.

12. See, for example, Ernst Snapper's chapter "The Three Crises in Mathematics: Logicism, Intuitionism, and Formalism," in *Mathematics: People, Problems, Results,* ed. Douglas Campbell and John Higgins (Belmont, Calif.: Wadsworth, 1984), vol. 2, p. 183; or Davis and Hersh, *The Mathematical Experience*, p. 319.

13. Davis and Hersh, *The Mathematical Experience*, p. 319.

14. Hardy, *A Mathematician's Apology*, p. 123.

15. Davis and Hersh, *The Mathematical Experience*, p. 321. They quote J. A. Dieudonne, who echoes this theme:

> On foundations we believe in the reality of mathematics, but of course when philosophers attack us with their paradoxes we rush to hide behind formalism and say, "Mathematics is just a combination of meaningless symbols." And then we bring out chapters 1 and 2 on set theory. Finally we are left in peace to go back to our mathematics and do it as we have always done, with the feeling each mathematician has that he is working with something real. This sensation is probably an illusion, but is very convenient. ("The Work of Nicholas Bourbaki," *American Mathematical Monthly* 77 [1970]: 145)

Because it is impossible to *prove* which mathematical philosophy is correct, mathematicians often avoid such discussions.

16. Hardy, *A Mathematician's Apology*, p. 121.

17. Russell, *My Philosophical Development*, p. 208.

18. David Bloor, *Knowledge and Social Imagery* (Chicago: University of Chicago Press, 1991), p. 50. Bloor also explores an important parallel between mathematics and religion which is relevant: the tradition of not questioning certain fundamental features of each. In religion, certain things are taken for granted, such as the existence of God. One does not ask questions such as "Who *is* this God?" "Why should I believe him?" One is supposed to take these things on faith. In mathematics, there is little support for questions such as "What *is* mathematics?" "Why is it important?" "What is its relationship to life?" Indeed, there is little room for self-reflection in general about mathematics and how it fits into a larger picture.

Bloor discusses why scientists in general are so reluctant to engage in a rigorous investigation of the nature of scientific knowledge, one that acknowledges sociological influences. He argues that this resistance can be illuminated by seeing the parallels to the sacred and profane in religion.

The puzzling attitude towards science would be explicable if it were being treated as sacred, and as such, something to be kept at a respectful distance. This is perhaps why its attributes are held to transcend and defy comparison with all that is not science but merely belief, prejudice, habit, error or confusion. The workings of science are then assumed to proceed from principles neither grounded in, nor comparable with, those operating in the profane world of politics and power. (Ibid., p. 47)

19. These ideas and arguments are developed in depth in Morris Kline's very interesting and engaging book *Mathematics, the Loss of Certainty* (New York: Oxford University Press, 1980).

20. Davis and Hersh, *The Mathematical Experience,* p. 347.

21. Sal Restivo, *Mathematics in Society and History* (Boston: Kluwer Academic Publishers, 1992), p. 115.

22. There are famous examples of theorems or proofs that were accepted as truths by the mathematics community and later were proven to be wrong—for example, the results obtained by Euler in the manipulation of infinite series. But for other kinds of examples, see discussions in Gian Carlo Rota, "The Concept of Mathematical Proof," Nicolas Goodman, "Modernizing the Philosophy of Mathematics," and other essays in *Essays in Humanistic Mathematics,* ed. Alvin White (Washington, D.C.: Mathematical Association of America, 1993).

23. As Kleiner says, "The axiomatic method in Greece did not come without costs. . . . Too much rigor may lead to rigor mortis." "Rigor and Proof in Mathematics," *Mathematics Magazine* 64, no. 5 (December 1991): 308.

24. Ibid., p. 301.

25. John Horgan, "The Death of Proof," *Scientific American* 269 (October 1993).

26. Ibid., p. 94.

27. Ibid., p. 100.

28. Davis and Hersh, *The Mathematical Experience,* p. 320.

29. Kleiner, "Rigor and Proof in Mathematics." There are sub-notes for the two quotes. The first is from J. Von Neumann, "The Role of Mathematics in the Sciences and in Society," in *Collected Works,* vol. 6, ed. A. H. Taub (New York: Macmillan, 1963), p. 480. The second is from H. Weyl, "Mathematics and Logic," *American Mathematical Monthly* 53 (1946): 13.

30. Horgan, "The Death of Proof," p. 100.

31. For a clear and elegant proof of this, see Ernest Nagel and James Newman, *Gödel's Proof* (New York: New York University Press, 1958).

32. Kline, *Mathematics, the Loss of Certainty,* p. 6.

33. Paul Halmos, "Mathematics as a Creative Art," in Campbell and Higgins, *Mathematics: People, Problems, Results,* vol. 2, p. 23.

34. Morris Kline, *Mathematics in Western Culture* (New York: Oxford University Press, 1953), p. 408.

35. The term *intuition* is usually used in one of two ways, though often people shift back and forth between meanings in the same conversation. On one hand, intuition is seen as a kind of innate instinct, while on the other hand it is seen as learned and hence something that can be developed. The vision of intuition as innate is linked to a belief in a biological basis for intuition that it is somehow rooted in the way our brains are formed. Idiot savants and child prodigies are seen as examples of people with highly developed intuition rooted in a physiological difference. The other kind of intuition is one that develops over time and is based

in experience. As Raymond Wilder writes, "Mathematical intuition, like intelligence, is a psychological quality which is . . . principally an accumulation of attitudes derived from one's mathematical experience." Moreover, "the more experienced the mathematician, the more reliable is his intuition." What both of these conceptions of intuition share, however, is a vision of intuition as something pre-rational, pre-formalized, and non-rigorous. It is the place of ideas, not verification. That is the way the term will be used here. Wilder, "The Role of Intuition," in *Mathematics: People, Problems, Results*, ed. Douglas Campbell and John Higgins, vol. 2 (Belmont, Calif.: Wadsworth, 1984).

36. James R. Newman, *The World of Mathematics* (New York: Simon and Schuster, 1956), vol. 3, pp. 1970–71.

37. Bloor, *Knowledge and Social Imagery*, p. 127. Boyer's quote is from *The History of Calculus and Its Conceptual Development* (New York: Dover, 1959), p. 169.

38. Bruce Berndt, "Srinivasa Ramanujan," *The American Scholar* 58, no. 2 (Spring 1989): 242.

39. "The Narrow Mathematician," *American Mathematical Monthly* 69 (June–July 1962): 463.

40. Russell, *My Philosophical Development*, p. 211.

41. White, *Essays in Humanistic Mathematics*, p. 97.

42. Restivo, *Mathematics in Society and History*, p. 45.

43. See, for example, ibid. Restivo discusses the use of Chinese ideographs and argues that they hindered mathematical development in some ways, and were used to maintain an exclusive access to intellectual pursuits by the elite. See also Marcia Ascher, *Ethnomathematics* (New York: Chapman and Hall, 1991), which discusses different directions that mathematics took in different cultures.

44. As David Goodstein describes in his article "Scientific Elites and Scientific Illiterates," *Engineering and Science* (Spring 1993), peer review is one of the "crucial pillars of the whole edifice" of scientific research. It is used to decide which papers will be published in scientific journals and who will receive grants from agencies such as the NSF. But there are significant limitations to peer review.

> Peer review is usually quite a good way of identifying valid science. Of course, a referee will occasionally fail to appreciate a truly visionary or revolutionary idea, but by and large peer review works pretty well so long as scientific validity is the only issue at stake. However, it is not at all suited to arbitrate an intense competition for research funds or for editorial space in prestigious journals. There are many reasons for this, not the least being the fact that the referees have an obvious conflict of interest, since they are themselves competitors for the same resources. It would take impossibly high ethical standards for referees to avoid taking advantage of their privileged anonymity to advance their own interests, but as time goes on, more and more referees have their ethical standards eroded as a consequence of having themselves been victimized by unfair reviews when they were authors. Peer review is thus one among many examples of practices that were well suited to the time of exponential expansion, but that will become increasingly dysfunctional in the difficult future we face." (P. 30)

45. Hardy, *A Mathematician's Apology*, p. 103.

46. Ibid., pp. 108–109.

47. Ibid., pp. 109–110.

48. Ibid., p. 112.

49. Hardy argues that this sense of aesthetic is one that both mathematicians and non-mathematicians can appreciate. He goes on to say, "It may be very hard to *define* mathematical beauty, but that is just as true of beauty of any kind—we may not know quite what we mean by a beautiful poem, but that does not prevent us from recognizing one when we read it." Ibid., p. 85.

50. Davis and Hersh, *The Mathematical Experience*, p. 168.

51. Ibid., p. 169.

52. Halmos, "Mathematics as a Creative Art," p. 28.

53. Eugene Wigner, "The Unreasonable Effectiveness of Mathematics in the Natural Sciences," *Communications on Pure and Applied Mathematics* 13 (1960): 1–14.

54. Kline, *Mathematics, the Loss of Certainty*, p. 6.

55. There has been a tremendous amount of change at the pre-college level in mathematics education in recent years, fueled largely by *Curriculum and Evaluation Standards for School Mathematics* (1989) and *Professional Standards for Teaching Mathematics* (1993), published by the National Council of Teachers of Mathematics. These reform movements stress the development of intuition and conjecture, and promote active and interactive learning by students. Change at the post-secondary level has been much slower.

56. Douglas Campbell, *The Whole Craft of Number* (Boston: Prindle, Weber and Schmidt, 1976). Quoted by Dorothy Buerk in "From Magic to Meaning: Changing the Learning of Mathematics," prepared for a Workshop on Teaching and Learning Mathematics, Augsburg College, St. Paul, Minn., 1985.

57. Johnetta Cole, *Conversations: Straight Talk with America's Sister President* (New York: Doubleday, 1993).

58. White, *Essays in Humanistic Mathematics*.

59. See, for example, Elizabeth Fennema and Gilah Leder, *Mathematics and Gender* (New York: Teachers College Press, 1990), p. 19, or Myra Sadker and David M. Sadker, *Failing at Fairness* (New York: Charles Scribner's Sons, 1994).

60. Pat Rogers, "Thoughts on Power and Pedagogy," in *Gender and Mathematics*, ed. Leone Burton (London: Cassell, 1990), pp. 42–43.

61. Mary Belenky, Blythe Clinchy, Nancy Goldberger, and Jill Tarule, *Women's Ways of Knowing* (Basic Books, 1986), p. 201.

62. Dorothy Buerk and Jackie Szablewski, "Getting Beneath the Mask, Moving Out of Silence," in White, *Essays in Humanistic Mathematics*.

63. Einstein, "Autobiographical Notes," p. 15.

64. Would women have a different perspective on what topics would be considered important, or what criteria to use in evaluating research? At this point, it is almost impossible to know if these kinds of decisions would be gender-related. For one, women are significantly underrepresented in the kinds of positions that shape future directions of mathematics—as editors of journals, as members of committees that grant awards and invite speakers, and as officers of funding agencies; hence their voices are less likely to be heard. Moreover, many of these judgments are communally shaped, and women are as likely to be influenced by their peers as their male counterparts, particularly in a field where being a woman can in some sense be seen as a liability, and hence there is even more pressure to conform.

65. Patricia Kenschaft and Sandra Keith, eds., *Winning Women into Mathematics* (Washington, D.C.: Mathematical Association of America, 1991).

66. Jerome Bruner, "On Learning Mathematics," *Mathematics Teacher* 53 (December 1960): 613.

Conclusion

1. Robert Edouard Moritz, *Memorabilia Mathematica: The Philomath's Quotation Book* (Washington, D.C.: Mathematical Association of America, 1914), p. 121.

RELEVANT LITERATURE

Women in Mathematics

COLLECTIVE BIOGRAPHIES

Campbell, Paul, and Louise Grinstein. *Women in Mathematics: A Biobibliographic Sourcebook.* Westport, Conn.: Greenwood Press, 1987.

Osen, Lynn. *Women in Mathematics.* Cambridge, Mass.: MIT Press, 1974.

Perl, Teri. *Math Equals.* Reading, Mass.: Addison Wesley, 1978.

INDIVIDUAL BIOGRAPHIES

Dick, August. *Emmy Noether, 1882–1935.* Boston: Birkhauser, 1981.

Koblitz, Ann. *A Convergence of Lives: Sofia Kovalevskaia—Scientist, Writer, Revolutionary.* Boston: Birkhauser, 1983.

Patterson, Elizabeth. *Mary Somerville, 1780–1872.* New York: Oxford University Press, 1979.

Stein, Dorothy. *Ada: A Life and a Legacy.* Cambridge, Mass.: MIT Press, 1985. (Biography of Ada Byron Lovelace)

OTHER

Keith, Sandra, and Philip Keith. *Proceedings of the National Conference on Women in Mathematics and the Sciences.* St. Cloud, Minn., 1990.

Kenschaft, Patricia, ed. *Winning Women into Mathematics.* Washington, D.C.: The Mathematical Association of America, 1991.

Mathematics Biography

Albers, Donald, and Gerald Alexanderson. *Mathematical People.* Boston: Birkhauser, 1985.

Albers, Donald; Gerald Alexanderson; and Constance Reid. *More Mathematical People.* New York: Harcourt Brace Jovanovich, 1990.

Bell, Eric Temple. *Men of Mathematics.* New York: Simon and Schuster, 1937, 1965.

Women in Science

Abir-Am, Pnina, and Dorinda Outram, eds. *Uneasy Careers and Intimate Lives: Women in Science, 1787–1979.* New Brunswick, N.J.: Rutgers University Press, 1987.

Alic, Margaret. *Hypatia's Heritage: A History of Women in Science from Antiquity to the Late Nineteenth Century.* Boston: Beacon Press, 1986.

Kass-Simon, G., and Patricia Farnes. *Women of Science: Righting the Record.* Bloomington: Indiana University Press, 1990.

Keller, Evelyn Fox. *A Feeling for the Organism: The Life and Work of Barbara McClintock.* New York: W. H. Freeman and Co., 1983.

McGrayne, Sharon. *Nobel Prize Women in Science.* New York: Birch Lane Press Book, published by Carol Publishing Group, 1993.

Mozans, H. J. *Women in Science.* New York: D. Appleton and Co., 1913. Reprint, Cambridge, Mass.: MIT Press, 1974.

Rossiter, Margaret. *Women Scientists in America: Struggles and Strategies to 1940.* Baltimore: Johns Hopkins University Press, 1982.

Schiebinger, Londa. *The Mind Has No Sex? Women in the Origins of Modern Science.* Cambridge, Mass.: Harvard University Press, 1989.

Zuckerman, Harriet; Jonathan Cole; and John Bruer, eds. *The Outer Circle: Women in the Scientific Community.* New York: W. W. Norton and Co., 1991.

Gender and Science

Birke, Lynda. *Women, Feminism, and Biology: The Feminist Challenge.* New York: Methuen, 1986.

Bleier, Ruth. *Science and Gender: A Critique of Biology and Its Theories on Women.* Elmsford, N.Y.: Pergamon, 1984.

Bleier, Ruth, ed. *Feminist Approaches to Science.* New York: Pergamon, 1986.

Fausto-Sterling, Anne. *Myths of Gender: Biological Theories about Women and Men.* New York: Basic Books, 1986.

Harding, Sandra. *The Science Question in Feminism.* Ithaca: Cornell University Press, 1986.

Harding, Sandra. *Whose Science? Whose Knowledge?* Ithaca, N.Y.: Cornell University Press, 1991.

Harding, Sandra, and Jean O'Barr, eds. *Sex and Scientific Inquiry.* Chicago: University of Chicago Press, 1987.

Keller, Evelyn Fox. *Reflections on Gender and Science.* New Haven: Yale University Press, 1985.

Longino, Helen. *Science as Social Knowledge.* Princeton, N.J.: Princeton University Press, 1990.

Lowe, Marian, and Ruth Hubbard, eds. *Woman's Nature: Rationalizations of Inequality.* New York: Pergamon, 1983.

Rose, Hilary. *Love, Power and Knowledge: Towards a Feminist Transformation of the Sciences.* Bloomington: Indiana University Press, 1994.

Tuana, Nancy. *Feminism and Science.* Bloomington: Indiana University Press, 1989.

Wertheim, Margaret. *Pythagoras' Trousers.* New York: Times Books, Random House, 1995.

Mathematics Education

Burton, Leone, ed. *Gender and Mathematics: An International Perspective.* Great Britain: Cassell, 1990.

Chipman, Susan; Lorelei Brush; and Donna Wilson, eds. *Women and Mathematics: Balancing the Equation.* Hillsdale, N.J.: Lawrence Erlbaum Associates, 1985.

Fennema, Elizabeth, and Gilah Leder, eds. *Mathematics and Gender*. New York: Teachers College Press, 1990.

Fox, Lynn; Linda Brody; and Dianne Tobin, eds. *Women and the Mathematical Mystique*. Baltimore: Johns Hopkins University Press, 1980.

Frankenstein, Marilyn. *Relearning Mathematics: A Different Third R—Radical Math(s)*. London: Free Association Books, 1989.

Rosser, Sue. *Female-Friendly Science*. New York: Pergamon Press, 1990.

Secada, Walter; Elizabeth Fennema; and Lisa Byrd Adajian, eds. *New Directions for Equity in Mathematics Education*. New York: Cambridge University Press (in collaboration with the NCTM), 1995.

Tobias, Sheila. *They're Not Dumb, They're Different: Stalking the Second Tier*. Tucson, Ariz.: The Research Corp., 1990.

Other Relevant Literature

Aisenberg, Nadya, and Mona Harrington. *Women of Academe: Outsiders in the Sacred Grove*. Amherst: University of Massachusetts Press, 1988.

Simeone, Angela. *Academic Women: Working towards Equality*. South Hadley, Mass.: Bergin and Garvey Publishers, 1987.

NAME INDEX

Adeboye, Adeniran, 224
Adleman, Leonard, 16
Adler, Alfred, 109
Agnesi, Maria, xxvii
Allen, Woody, 44
Allendoerfer, Carl, 249
Archibald, Nate (Tiny), 228

Baker, Ella, 194
Baxter, Arlene, 144
Bellman, Richard, 220
Bird, Larry, 228
Birkhoff, George, 234
Birman, Joan, xi, 21, 81, 112, 114, 117,
    120–140, 158, 270, 271
Blackwell, David, 230
Bloor, David, 240, 248, 281
Blum, Lenore, xii, xv, 73, 77, 80, 82, 114,
    142, 144–164, 174, 260, 270, 271, 279
Bollobas, Bela, 116
Bourguignon, J. P., 35
Boyer, Carl, 248
Brown, Marjorie Lee, 280
Buerk, Dorothy, 258
Burton, Leone, 280

Campbell, Douglas, 256
Carroll, Constance, 208
Cartwright, Mary, 116
Chachere, Gerald, 224
Choquet-Bruhat, Yvonne, xxvi
Chung, Fan, xii, 11, 16, 20, 21, 73, 80, 82,
    83, 90, 96–108, 112, 270, 275, 276
Clarke, Edward C., xxv
Cole, Johnetta, 189, 257
Copes, Larry, 258

De Morgan, Augustus, xxv, 4
Denamark, Freida, 215
Diaconis, Persi, 108
Donaldson, Jim, 223–224
Donaldson, Simon, 46
Du Bois, W. E. B., 194, 195, 199

Einstein, Albert, 62, 235, 261

Erdos, Paul, 16
Erving, Julius (Dr. J), 228
Ewing, Patrick, 228

Fleissner, Bill, 177
Fox, Mary Frank, 17
Fox, Ralph, 128, 131, 132, 270
Freed, Daniel, 46
Freire, Paulo, 257

Gauss, Carl Friedrich, xxvi, 116, 117
Geller, Sue, 274
Germain, Sophie, xxv, 267
Gill, Tepper, 224
Gödel, Kurt, xiv, 56, 62, 246
Goodstein, David, 283
Gorenstein, Daniel, 244
Graham, Ron, 80, 97, 100–104, 108
Granville, Evelyn Boyd, 10, 200–202, 209,
    280
Gray, Mary, 152, 279
Groszek, Marcia, xiii, 10, 16, 17, 76, 81, 274

Hahn, Hans, 247, 248
Halmos, Paul, 247, 254, 267, 271
Hardy, G. H., 71, 106, 107, 109, 236, 239,
    240, 252, 253, 284
Harrison, Jenny, 277
Hay, Louise, 34
Helson, Ravenna, xxxi, 7, 91, 262
Hilbert, David, 271, 277
Hilgendorf, Lila, 90
Hirsch, Moe, 279
Hoppensteadt, Frank, 222
Horgan, John, 244
Hunt, Fern, xiii, 6, 7, 10, 11, 13, 78, 212–
    233
Hutchinson, Joanne, 100

Johnson, Magic, 228
Jones, Vaughan, 139, 140
Jordan, Camille, 5
Jordan, Michael, 228

Kanimori, Aki, 16

Keen, Linda, xiii, 11
Keller, Evelyn Fox, 183
Keller, Helen, 227
Kenschaft, Patricia, 75, 261
Keyser, Cassius, 236
King, Martin Luther, Jr., 194, 210
Klein, Felix, 271
Kleiner, Israel, 243, 245
Kline, Morris, 246, 247
Koblitz, Ann, 267
Kovalevskaia, Sofia, xxv, 68, 69, 91, 267
Kreinberg, Nancy, 154, 155
Kunen, Ken, 70, 174

Leder, Gilah, 8
Lee, Alton B., 194
Lorch, Lee, 200–202, 209
Lovelace, Lady Byron, xxv, 4
Lyttle, George, 121

MacIntyre, Diane, 279
Magnus, Wilhelm, 127, 130, 132
Malcolm X, 194
Malone, P. R., 195–198
Malone-Mayes, Vivienne, xv, 10, 12, 13, 78,
    189, 192–211, 214
Marshall, Durkin, 200
Marx, Karl, 162
Mayes, J. J., 204
Meitner, Lise, xxvii
Moore, A. D., 219
Moore, R. L., 11, 14, 87–91
Morawetz, Cathleen, 71, 74, 217
Moses, Bob, 189
Mumford, David, 108
Murphy, Larry, 24

Neumann, J. von, 245
Nirenberg, Louis, 129
Noether, Emmy, 68, 91, 267
Northrop, E. P., 236
Nyikos, Peter, 177

Orey, Steven, 57
Osen, Lynn, 267

Palais, Dick, 30
Parks, Rosa, 193
Perl, Teri, 267
Peter, Rozsa, xx

Pierpont, James, 4
Pippen, Scottie, 228
Pless, Vera, 34
Pour-El, Akiva, 55, 58, 61, 62
Pour-El, Marian, xiv, 48–64, 70, 73, 75, 81,
    81–83, 112, 272, 273
Pythagoras, 116, 253

Ramanujan, Srinivasa, 248
Rhodes, John, 279
Richards, Ian, 57
Richardson, Sid, xvi, 39
Robinson, Abraham, 149, 151
Robinson, Julia, 150, 164, 174, 279
Rogers, Pat, 258
Roitman, Judy, xv, 7, 10, 12, 70, 74, 76, 77,
    81, 112, 113, 142, 153, 166–187, 270,
    273
Rossiter, Margaret, xviii, xxvii, 145
Rota, Gian Carlo, 251, 278
Rudin, Mary Ellen, xv, 6, 11, 14, 80, 82–94,
    113, 114, 117, 142, 167, 174, 175, 179,
    183, 185, 267, 270, 275, 278
Rudin, Walter, 80, 89–92, 160, 275
Russell, Bertrand, 236, 240, 243, 250
Russell, Bill, 228

Sacks, Gerald, 16, 149, 150
Sacks, Jonathan, 35
Satter, Ruth Lyttle, 122
Schafer, Alice, 279
Schoen, Rick, 35
Sells, Lucy, xxix, 279
Serge, Corrado, 19
Shub, Mike, 159, 164, 279
Simon, Leon, 35
Singer, I. M., 148
Smale, Steven, 159, 164, 279
Smith, D. E., 116
Snyder, Martin Avery, 221
Solow, Anita, x
Steen, Lynn Arthur, 263
Stern, Nancy, 112–113
Sternberg, Shlomo, 108
Sweatt, Hermann Marion, 209–210
Sylvester, J. J., 116
Synge, J. L., 235
Szego, Gabor, 271

Taub, Abe, 272

Taubes, Cliff, 35
Thomas, Isiah, 228
Thomas, J. M., 89
Thurston, William, 139, 140, 184, 246
Tobias, Sheila, 156
Treisman, Uri, 279

Uhlenbeck, Karen, xvi, 7, 13, 15, 24, 25–46, 77, 81, 115, 118, 175, 269, 270, 272
Uhlenbeck, Olke, 31
Ulam, Stan, 278

Walton, Bill, 228
Washington, Booker T., 194, 195, 199
Weidman, Donald, 5
Weierstrass, Karl, xxv, 4, 243, 265

Weil, Andre, 110
Weyl, Hermann, 68, 245, 255
Whitehead, Alfred North, 78, 243
Whitney, Hassler, 53
Wiegand, Sylvia, 110
Wilder, R. L., 89
Wiles, Andrew, 5, 9, 17, 21, 244
Wilf, Herbert, 100, 275
Wilson, Charles, 216–217
Wilson, Woodrow, 29
Witten, Ed, 42

Yau, Shing Tung, 35, 37, 108, 270
Young, Grace Chisolm, xxv, xxvii, 91, 110, 267
Young, William, xxvii

CLAUDIA HENRION is a Visiting Professor at Dartmouth College. She teaches courses in mathematics, education, and the Master of Arts and Liberal Studies Program. She taught mathematics at Middlebury College for seven years.